Biometrics and Cryptography

Edited by Sudhakar Radhakrishnan and Carlos M. Travieso-González

Published in London, United Kingdom

Biometrics and Cryptography
http://dx.doi.org/10.5772/intechopen.107597
Edited by Sudhakar Radhakrishnan and Carlos M. Travieso-González

Contributors
Asma'a Al-Hakimi, Bernadette Dorizzi, Carlos M. Travieso-Gonzalez, Dijana Petrovska-Delacretaz, El Hadj Youssef Wajih, Gaurav Verma, Gideon Samid, José Luis Vásquez-Vasquez, Menachem Domb, Misbah ul Iman, Mohamed Amine Hmani, Monther Tarawneh, Muhammad Ibrahim Ravi Bin Gobi, Sherine Jenny Rajan, Sudhakar Radhakrishnan, Wenqi He, Xiang Peng

First published in London, United Kingdom, 2024 by IntechOpen
IntechOpen is the global imprint of INTECHOPEN LIMITED, registered in England and Wales, registration number: 11086078, 167-169 Great Portland Street, London, W1W 5PF, United Kingdom

British Library Cataloguing-in-Publication Data
A catalogue record for this book is available from the British Library

Additional hard and PDF copies can be obtained from orders@intechopen.com

Biometrics and Cryptography
Edited by Sudhakar Radhakrishnan and Carlos M. Travieso-González
p. cm.
Print ISBN 978-1-83768-262-1
Online ISBN 978-1-83768-263-8
eBook (PDF) ISBN 978-1-83768-264-5

For EU product safety concerns:
IN TECH d.o.o., Prolaz Marije Krucifikse Kozulić 3, 51000 Rijeka, Croatia,
info@intechopen.com or visit our website at intechopen.com.

Meet the editors

Dr. Sudhakar Radhakrishnan is currently a professor and the head of the Department of Electronics and Communication Engineering, Dr. Mahalingam College of Engineering and Technology, Pollachi, India. He is currently an associate editor of *IEEE-Access* and *IET-Image Processing*, published by IEEE and Wiley, respectively. He got the Outstanding Associate Editor award for IEEE Access in the year 2019. He is a reviewer of 16 international journals, namely *IEEE Transactions on Systems, Man, and Cybernetics: Systems by IEEE*; *International Arab Journal of Information Technology* coming from Zarqa University, Jordan; *International Journal of Computer and Electrical Engineering* published by the International Association of Computer Science and Information Technology Press (IACSIT); *International Journal of Computer Theory and Engineering* published by the International Association of Computer Science and Information Technology Press (IACSIT); *Journal of Electrical and Electronics Engineering Research*; *Iranian Journal of Electrical and Computer Engineering* published by the Iranian Research Institute for Electrical Engineering, JD, Iran; *Journal of Optical Engineering* published by SPIE, USA; *Journal of Electronic Imaging* published by SPIE, USA; *Imaging Science Journal*, published by Taylor & Francis Group; *International Journal of Computational Science and Engineering (IJCSE)*, published by Inderscience; *International Journal of Image Mining* (IJIM) published by Inderscience; *International Journal of Biomedical Engineering and Technology* (IJBET) published by Inderscience; *Journal for Image Analysis & Stereology* from the International Society for Stereology; *ETRI Journal* from Korea; *AEUE—International Journal of Electronics and Communications* by Elsevier; and *IET Image Processing*. He wrote two books titled, *Research Issues in Image Compression Using Wavelet Variants* (Lambert Academic Publishing [LAP], Germany,.2010) and *Practicing Signals and Systems Laboratory Using MATLAB* LAP, Germany, 2010) and two book chapters titled "Wavelet Based Image Compression" in *Computational Intelligence Techniques in Handling Image Processing and Pattern Recognition* (LAP, Germany (2010) and "Analysis of Hand Vein Images Using Hybrid Techniques" in *Hybrid Intelligence Techniques for Image Analysis and Understanding* (Wiley). He edited six books under IntechOpen Publisher, Croatia. He has published 140 papers in international and national journals and conference proceedings. His areas of research include digital image processing, image analysis, wavelet transforms, and digital signal processing.

Carlos M. Travieso-González received an M.Sc. degree in 1997 in telecommunication engineering at the Polytechnic University of Catalonia (UPC), Spain, and a Ph.D. degree in 2002 at the University of Las Palmas de Gran Canaria (ULPGC-Spain). He is a full professor of signal processing and pattern recognition and is the head of the Signals and Communications Department at ULPGC, teaching from 2001 on subjects on signal processing and learning theory. His research lines are biometrics, biomedical signals and images, data mining, classification systems, signal and image processing, machine learning, and environmental intelligence. He has researched 54 international and Spanish research projects, some of them as head researcher. He is co-author of four books, co-editor of 25 proceedings books, and guest editor for eight JCR-ISI international journals and up to 24 book chapters. He has over 480 papers published in international journals and conferences (95 of them indexed on JCR–ISI–Web of Science). He has published seven patents in the Spanish Patent and Trademark Office. He has been a supervisor on 12 Ph.D. theses (with seven more under supervision) and 150 masters theses. He is the founder of the IEEE IWOBI conference series and president of its Steering Committee, the InnoEducaTIC conference series, and the APPIS conference series. He is an evaluator of project proposals for the European Union (Horizon Europe and H2020—EU), Medical Research Council (MRC—UK), Spanish government (ANECA—Spain), Spanish Research Agency (AEI—Spain), Croatian Science Foundation (HRZZ—Croatia), Research National Agency (ANR—France), DAAD in Spain (Germany), the Argentinian government, and Colombian institutions. He has been a reviewer in different indexed international journals (<75) and conferences (<250) since 2001. He has been a member of the IASTED Technical Committee on Image Processing from 2007 and a member of the IASTED Technical Committee on Artificial Intelligence and Expert Systems from 2011. He has been APPIS 2018, 2019, and2022–2024 general chair; IEEE-IWOBI 2014, 2015, 2017, 2019, and 2022 general chair, ACM-APPIS 2020 general chair, InnoEducaTIC 2014 and 2017 general chair; IEEE-INES 2013 general chair, NoLISP 2011 general chair, JRBP 2012 general chair, and IEEE-ICCST 2005 co-chair. He is an associate editor of the *Computational Intelligence* and *Neuroscience* journal (Hindawi—Q2 JCR-ISI) in entropy (MDPI—Q2 JCR-ISI) and sensors (MDPI—Q1 JCR-ISI). He was vice-dean from 2004 to 2010 in the Higher Technical School of Telecommunication Engineers at ULPGC, and vice-dean of Graduate and Postgraduate Studies from March 2013 to November 2017. He won the "Catedra Telefonica" Awards in the modality of Knowledge Transfer in 2017, 2018, and 2019; in 2020, in the modality of COVID Research; and in 2021, 2022, and 2023 in the modality of Research Project. He received the Spanish national award in the second edition of *E-nnova Health* 2022 of *Diario Médico* and *Correo Farmacéutico* in the category of Big Data and Artificial Intelligence, for the proposal entitled "Implementation of the digital twin in SARS-COV-2 patient as a prognostic prediction," in October 2022. He was a finalist in the ABB Award in 2023 with a proposal entitled "Energy Predictor" in the modality of public administration. In 2022, he was one of the researchers included as a member of the 2 percent most influential group in the world, according to the 2022 edition of the Stanford ranking compiled by John P. A. Ioannidis (Stanford University), Kevin W. Boyack (SciTech Strategies), and Jero en Baas (Research Intelligence, Elsevier B.V.), which analyses his scientific output and the citations he has received, among other criteria.

Contents

Preface IX

Section 1
Cryptography 1

Chapter 1 3
Introductory Chapter: Exploring Cryptography – Examining Its Fundamentals and Potential Applications
by Sudhakar Radhakrishnan and Sherine Jenny Rajan

Chapter 2 9
Perspective Chapter: Cryptography – Recent Advances and Research Perspectives
by Monther Tarawneh

Chapter 3 39
Securing Java Source Files Using the Cipher Cryptographic Algorithm
by Asma'a Al-Hakimi, Muhammad Ibrahim Ravi Bin Gobi and Misbah ul Iman

Chapter 4 61
Pattern Devoid Cryptography
by Gideon Samid

Section 2
Biometrics 85

Chapter 5 87
Introductory Chapter: The Fundamentals of Biometrics – A Comprehensive Exploration
by Carlos M. Travieso-González

Chapter 6 91
Revocable Crypto-Biometric Key Regeneration Using Face Biometrics, Fuzzy Commitment, and a Cohort Bit Selection Process
by Mohamed Amine Hmani, Dijana Petrovska-Delacretaz and Bernadette Dorizzi

Chapter 7 **115**
Advanced Lightweight Encryption Key Management Algorithms for IoT Networks
by Menachem Domb

Chapter 8 **129**
Artificial Neural Network Computational Techniques in Biometric Handwriting
by Jose Luis Vásquez-Vasquez and Carlos M. Travieso-González

Chapter 9 **153**
Biometric-Based Optical Systems for Security and Authentication
by Gaurav Verma, Wenqi He and Xiang Peng

Chapter 10 **177**
Secure Smart Card IP
by El Hadj Youssef Wajih

Preface

Fast communication between the digital and physical worlds is made possible by the Internet of Things (IoT). This rapid growth has made everything in the world able to get connected. Therefore, securing this growing amount of data while it is being transmitted will be essential. Cryptography is a crucial instrument used to protect this data.

The study of encrypting and decrypting data using mathematics is known as cryptography. With cryptography, you can send or keep confidential data over unsecured networks, such as the Internet, so that only the intended receiver can read it. In our IoT-connected world, cryptography is utilized to encrypt all transferred data and authenticate individuals, devices, and other gadgets. The study of data security is known as cryptography, and the science of deciphering and intercepting secure communication is known as cryptanalysis. Cryptography and cryptanalysis together are said to be cryptology.

Cryptography has crept into everything, from Web browsers and e-mail programs to cell phones, bank cards, and cars. Soon, we will see many new exciting applications for cryptography such as radio frequency identification (RFID) tags for anti-counterfeiting. As a consequence of the pervasiveness of crypto algorithms, an increasing number of people must understand how they work and how they can be applied in practice. This book addresses this issue by providing a comprehensive introduction to modern applied cryptography that is equally suited for students and practitioners in industry.

Much of the focus is on practical relevance by introducing most crypto algorithms that are used in modern real-world applications. In addition to crypto algorithms, attention is also given to important cryptographic protocols, modes of operation, security services, and key establishment techniques. Timely topics include lightweight ciphers, which are optimized for constrained applications.

The chapters in the first section of the book explore the latest developments, methods, approaches, and applications of cryptography in a wide variety of fields and endeavors. This section provides researchers, academicians, and readers of various backgrounds and methods with an in-depth discussion of the latest advances. It consists of 4 chapters from academicians, practitioners, and researchers in the field of cryptography.

In the second section, the book discusses biometrics. Biometrics is a fascinating discipline that uses unique characteristics of the human body for the identification and verification of individuals. This section explores the diverse applications of biometrics, from unlocking mobile phones to access control in high-security facilities. Through practical examples and clear explanations, we will discover how biometrics is transforming the way we interact with technology and how it ensures greater security in our society. Welcome to a world where your body is your most secure identification.

Dr. Sudhakar Radhakrishnan
Professor,
Department of Electronics and Communication Engineering,
Dr. Mahalingam College of Engineering and Technology,
Tamil Nadu, India

Dr. Carlos M. Travieso-González
Professor,
Department of Signals and Communication,
Institute for Technological Development and Innovation in Communications,
University of Las Palmas,
Gran Canaria, Spain

Section 1

Cryptography

Chapter 1

Introductory Chapter: Exploring Cryptography – Examining Its Fundamentals and Potential Applications

Sudhakar Radhakrishnan and Sherine Jenny Rajan

1. Introduction

Fast communication between the digital and physical worlds is made possible by the Internet of Things (IoT).

This rapid growth has made everything in the world to get connected. Therefore, it will be essential to secure this growing amount of data while it is being transmitted. Cryptography is a crucial instrument used to protect this data [1].

The study of encrypting and decrypting data using mathematics is known as cryptography. With the use of cryptography, you can send or keep confidential data over unsecure networks, such as the Internet, such that only the intended receiver can read it. In our IoT-connected world, cryptography is utilized to encrypt all transferred data as well as to authenticate individuals, devices, and other gadgets. The study of data security is known as cryptography, and the science of deciphering and intercepting secure communication is known as cryptanalysis. Cryptography and Cryptanalysis together is said to be Cryptology. The history of cryptography is interesting and extraordinarily long and may be found in the book "The Code Book: The Secrets Behind Code breaking" by Simon Singh.

2. Fundamentals

The goals of cryptography are confidentiality, integrity and availability. Information confidentiality is safeguarded by cryptography, which also makes sure data has not been altered and that the information originated from the intended source rather than an unauthorized sender [2].

Symmetric key and asymmetric key cryptography are the two forms of cryptography. Using the same secret key, the sender and the recipient encrypt and decode a message in symmetric cryptography [3]. Refer **Figure 1**. Most widely used AES belongs to symmetric key cryptography. Owing to its less complex mathematical procedures, symmetric key cryptography is typically quicker and more effective than asymmetric encryption. This makes it appropriate for real-time encryption of massive amounts of data, like that needed to secure network connections and data storage.

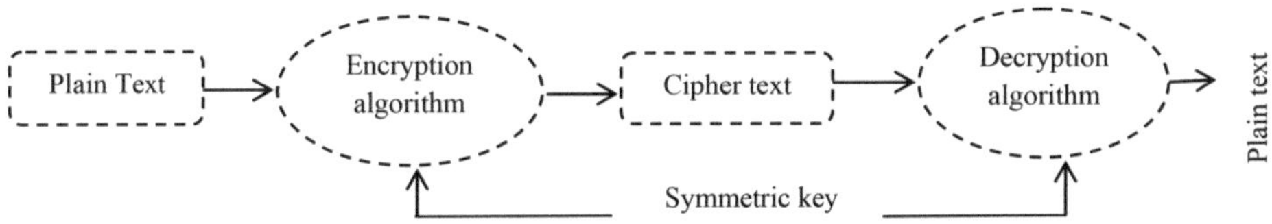

Figure 1.
Symmetric encryption.

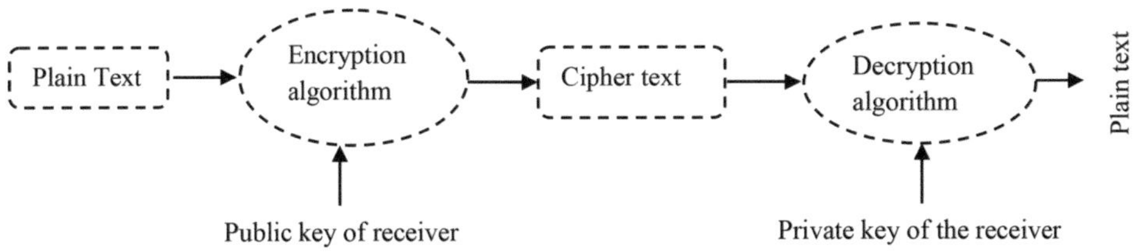

Figure 2.
Asymmetric encryption.

A pair of keys, private and public key are used by Asymmetric key cryptography. A few of the algorithms are Diffie-Helman key exchange, RSA and others. The private key is kept private and is only known to its owner, but the public key is freely shared and can be circulated broadly [4]. To decrypt messages encrypted with a public key, only the corresponding private key is needed, and vice versa. Digital signatures, the cryptographic equivalent of handwritten signatures, are made possible via asymmetric cryptography [5]. A message or document can be signed by the sender using their private key, and the recipient can verify the authenticity of the signature using the sender's public key. The message's origin and integrity are guaranteed by this procedure. The public key can be provided by a trusted third party like certificate authority (CA). Asymmetric encryption is exemplified in **Figure 2**, where the recipient's public key is used to encrypt the plain text, and the recipient's private key is used to crack it. This provides confidentiality.

3. Challenges and research avenue

The art of secure communication, cryptography, has several theoretical and practical difficulties [6]. Among the major challenges are

- Key management: the hardest part of securing large-scale systems is frequently managing the safe distribution, storage, and frequent updating (even during a single session) of secret keys [7].

- Computational power: attacks on standard encryption techniques can become more likely as computational power rises. The constant development of more robust algorithms that can survive ever-powerful computers is required of cryptographers [8].

- Side-channel attacks: it takes into account adversaries attempting to exploit the physical characteristics of real cryptography equipment [9]. Algorithm vulnerabilities provide a serious threat to the security of encrypted data.

- Algorithm vulnerabilities: vulnerabilities in even well-designed algorithms may not always be obvious at first [10]. A security vulnerability can expose all sensitive information to the attacker.

- Threats from quantum computing: quantum computers can solve some mathematical problems far quicker than conventional computers. Their ability to overcome well-known encryption methods like RSA and ECC is considerable [11].

Numerous avenues for research into cryptography have been made possible by the difficulties in creating cryptographic algorithms. A great deal of recent effort has gone into creating post-quantum encryption that is resistant to quantum attacks, aiming to develop workable quantum-resistant cryptographic schemes. Current research examines methods such as homomorphic encryption to perform computations on encrypted data without exposing it. Research concentrating on standardization efforts to promote interoperability while maintaining security is in process. Further research on designing algorithms and implementations resistant to side channel attacks and optimizing cryptographic algorithms and protocols for better performance without compromising security is carried out.

4. Applications

In many real-time applications, cryptography is essential for maintaining data confidentiality and privacy on a variety of digital platforms. Cryptography protects sensitive data in the domain of online banking and financial transactions by using encryption algorithms to prevent theft or unauthorized access to user passwords [12], credit card information, and transactional data. Cryptographic protocols like SSL/TLS are used by secure communication channels, such email services and instant messaging applications, to encrypt data during transmission, protecting attachments and messages from tampering or eavesdropping. Cryptographic techniques are also used in the healthcare industry to secure electronic health records (EHRs), protecting patient privacy and adhering to legal requirements. Furthermore, encryption plays a major role in the developing field of block chain technology [13] by guaranteeing the immutability and integrity of distributed ledgers, allowing safe transactions. Many real time applications like Web Browsing [14], Messaging Apps [15], Bitcoin Transactions [13], Full Disk Encryption [16], Cloud Storage Encryption [17], Two-Factor authentication [12], Digital Signatures in Emails, Smart Home Devices & IoT devices [1], Military, Document security has increased the significance of encryption.

5. Conclusion

Ultimately, cryptography ought to be viewed as the fundamental component of digital security, a vital tool for safeguarding personal information and ensuring secure communication in a range of settings. An overview of the basic ideas, challenges and applications of cryptographic systems has been given in this introductory chapter. Next chapters will explore the intricacies of encryption algorithms, key management, and real-world applications of cryptography in a variety of sectors as we continue our investigation of this fascinating field. Moreover, the ever evolving technological landscape poses a constant threat to cryptographic systems, necessitating constant innovation and adaptation to counter new attacks. By carefully examining its roots and applications, this study aims to shed light on how cryptography supports data integrity, confidentiality, and authenticity in our increasingly interconnected world.

Author details

Sudhakar Radhakrishnan* and Sherine Jenny Rajan
Dr. Mahalingam College of Engineering and Technology, Coimbatore Dt, Tamil Nadu, India

*Address all correspondence to: sudha_radha2000@yahoo.co.in

References

[1] Fursan Thabit GHA-G, Can O, Aljahdali AO. Cryptography algorithms for enhancing IoT security. Internet of Things. 2023;**22**. Article No. 100759. DOI: 10.1016/j.iot.2023.100759

[2] Wu C-K. Fundamentals of cryptography. In: Internet of Things Security. Advances in Computer Science and Technology. Singapore: Springer; 2021

[3] Kumar T, Wollinger S. Fundamentals of symmetric cryptography. In: Lemke M, Paar K, Wolf C, editors. Embedded Security in Cars. Berlin, Heidelberg: Springer; 2006. pp. 125-143

[4] Diffie W, Hellman ME. New directions in cryptography. IEEE Transactions on Information Theory. 1976;**22**(6):644-654. DOI: 10.1109/TIT.1976.1055638

[5] Rivest RL, Shamir A, Adleman L. A method for obtaining digital signatures and public-key cryptosystems. Communications of the ACM. 1978;**21**(2):120-126. DOI: 10.1145/359340.359342

[6] Menezes A, Stebila D. Challenges in cryptography. IEEE Security and Privacy. 2021;**19**(2):70-73. DOI: 10.1109/MSEC.2021.3049730

[7] Shahnawaz Ahmad SMB. Hybrid cryptographic approach to enhance the mode of key management system in cloud environment. The Journal of Supercomputing. 2023;**79**:7377-7413. DOI: 10.1007/s11227-022-04964-9

[8] Bhat KJG, Iqbal M. Impact of computational power on cryptography. In: Giri KJ, Parah SA, Bashir R, Muhammad K, editors. Multimedia Security. Algorithms for Intelligent Systems. Singapore: Springer; 2021

[9] Standaert F-X. Introduction to side-channel attacks. In: Verbauwhede I, editor. Secure Integrated Circuits and Systems. Integrated Circuits and Systems. Boston, MA: Springer; 2010

[10] Zhou Y, Ma F, Chen Y, Ren M, Jiang Y. CLFuzz: Vulnerability detection of cryptographic algorithm implementation via semantic-aware fuzzing. ACM Transactions on Software Engineering and Methodology. 2023;**33**(2):1-28. DOI: 10.1145/3628160

[11] Renner R, Wolf R. Quantum advantage in cryptography. AIAA Journal. 2023;**61**(5):1895-1910. DOI: 10.2514/1.J062267

[12] Ometov A, Bezzateev S, Mäkitalo N, Andreev S, Mikkonen T, Koucheryavy Y. Multi-factor authentication: A survey. Cryptography. 2018;**2**(1):1-31. DOI: 10.3390/cryptography2010001

[13] Zhai S, Yang Y, Li J, Qiu C, Zhao J. Research on the application of cryptography on the blockchain. Journal of Physics Conference Series. 2019;**1168**(3):1-8. DOI: 10.1088/1742-6596/1168/3/032077

[14] Trevisan M, Soro F, Mellia M, Drago I, Morla R. Does domain name encryption increase users' privacy? ACM SIGCOMM Computer Communication Review. 2020;**50**(3):16-22. DOI: 10.1145/3411740.3411743

[15] Trauthig IK, Martin ZC, Woolley SC. Messaging apps: A rising tool for informational autocrats. Political Research Quarterly. 2023:1-13. DOI: 10.1177/10659129231190932

[16] Sassani BA, Alkorbi M, Jamil N, Naeem MA, Mirza F. Evaluating

encryption algorithms for sensitive data using different storage devices. Scientific Programming. 2020;**2020**:1-9. DOI: 10.1155/2020/6132312

[17] Mahato GK, Chakraborty SK. A comparative review on homomorphic encryption for cloud security. IETE Journal of Research. 2023;**69**(8): 5124-5133. DOI: 10.1080/03772063.2021.1965918

Chapter 2

Perspective Chapter: Cryptography – Recent Advances and Research Perspectives

Monther Tarawneh

Abstract

Cryptography is considered as a branch of both mathematics and computer science, and it is related closely to information security. This chapter explores the earliest known cryptographic methods, including the scytale, Caesar cipher, substitution ciphers, and transposition ciphers. Also, explains the evolution of these methods over time. The development of symmetric and asymmetric key cryptography, hash functions, and digital signatures is also discussed. The chapter highlights major historical events and technological advancements that have driven the need for stronger and more efficient encryption methods. In addition, the chapter explores the potential for integrating artificial intelligence tools with cryptographic algorithms and the future of encryption technology.

Keywords: cryptography, mathematics, computer science, information security, scytale, Caesar cipher, substitution ciphers, transposition ciphers, symmetric key cryptography, asymmetric key cryptography, hash functions, digital signatures, evolution, historical events, technological advancements, artificial intelligence, future

1. Introduction

Cryptography is the science converting information into an unreadable format as a practice of protecting confidential messages from unauthorized access [1]. Cryptographic algorithms have come a long way since the early days of cryptography and have evolved to keep up with the changing technological landscape. In this chapter, we will explore the history of cryptographic algorithms and their evolution over time.

The earliest known cryptographic methods date back to ancient civilizations, where methods, such as simple substitution and transposition ciphers, were used to conceal messages and prevent non-authorized people from understanding messages. These methods evolved over time to include more complex ciphers, such as the Caesar cipher and the Vigenère cipher, which were used during the Middle Ages. The development of the printing press and the subsequent increase in literacy rates led to the need for more secure methods of encryption, which led to the development of more complex ciphers such as the Playfair cipher and the Enigma machine.

Symmetric key cryptography is one of the oldest and most widely used types of encryption. It is based on the concept of using the same key to encrypt and decrypt a message. The history of symmetric key algorithms dates back to ancient times, where simple substitution ciphers were used to encrypt messages. Over time, more complex algorithms were developed such as the Hill cipher and the data encryption standard (DES). The development of the advanced encryption standard (AES) in the late twentieth century marked a significant improvement in symmetric key cryptography as it provided stronger encryption and faster processing times.

Asymmetric key cryptography, also known as public-key cryptography, is a more recent development in the field of cryptography. It is based on the use of two different keys—a public key and a private key—to encrypt and decrypt messages. The concept of asymmetric key cryptography was first introduced by Whitfield Diffie and Martin Hellman in 1976 [2]. This led to the development of various algorithms such as the Rivest-Shamir-Adleman (RSA) algorithm [3] and the Diffie-Hellman key exchange [4].

Hash functions are another important component of modern-day encryption. A hash function is a mathematical function that takes an input (or message) and produces a fixed-length output (or hash) [5]. Hash functions are used to ensure the integrity of data as any change to the original input will result in a different hash. The history of hash functions dates back to the 1950s, where the concept of message digests was introduced. Over time, more complex algorithms were developed such as the secure hash algorithm (SHA) and the message digest (MD) [5, 6].

Digital signatures are used to provide authentication and non-repudiation in digital communications. A digital signature is a mathematical scheme for demonstrating the authenticity of a digital message or document. The history of digital signature algorithms dates back to the early 1980s, where the concept of public-key cryptography was first introduced. Over time, various algorithms were developed such as the digital signature algorithm (DSA) and the elliptic curve digital signature algorithm (ECDSA) [7].

The evolution of cryptographic algorithms has been driven by major historical events and technological advancements. With the advent of the internet and the increase in digital communication, the need for stronger and more efficient encryption methods became more pressing. As computing power continues to increase, the potential for cracking encryption algorithms also increases. This has led to the need for stronger and more advanced cryptographic algorithms, such as post-quantum cryptography, which can withstand attacks from quantum computers.

In addition to the potential threats to encryption technology, there is also the potential for integrating artificial intelligence tools with cryptographic algorithms. For example, machine learning algorithms could be used to identify potential vulnerabilities in encryption systems and improve their security.

As the digital landscape continues to evolve, the importance of staying ahead of the curve in encryption technology cannot be overstated. This chapter provides an overview of the history and evolution of cryptographic algorithms, highlighting the need for ongoing innovation and development in this field. By continuing to push the boundaries of encryption technology, we can help to safeguard the privacy and security of sensitive data in the digital age.

Encryption is a critical component of modern communication and information security [8]. By converting data into a secure format that can only be accessed with the correct key or password, encryption ensures that sensitive information is protected from unauthorized access. Throughout history, cryptography has played a significant role in the security of sensitive information from the early substitution ciphers used by ancient civilizations to the modern public-key encryption algorithms.

Recent developments in technology have led to new challenges and opportunities in the field of cryptography. The rise of quantum computing [9], blockchain technology [10], and the need for secure communication in an increasingly connected world have all driven new research and innovation in the field of cryptography [11].

This chapter provides an overview of various cryptographic techniques, including symmetric and asymmetric encryption, hashing, digital signatures, homomorphic encryption, multiparty computation, and lightweight cryptography. Each of these techniques has its own strengths and weaknesses and is suited to different use cases and scenarios. The chapter also explores the future of cryptography, including developments in post-quantum cryptography, blockchain-based cryptography, and other emerging technologies. By understanding the principles and applications of modern cryptography, we can better protect our digital assets and maintain the privacy and security of our communication.

2. Ancient cryptography methods

The history of cryptography dates back to ancient civilizations, where people used various methods to protect their messages from unauthorized access. The earliest examples of cryptography being used to protect information were found in an inscription carved around 1900 BC, in the main chamber of the tomb of the nobleman Khnumhotep II, in Egypt [12, 13]. The inscription, known as the "Cryptography Inscription," described a method for hiding the meaning of hieroglyphic inscriptions by using symbols to represent individual letters. The symbols were then scrambled in a specific way to make the text difficult to read. The main purpose of the "Cryptography Inscription" was not to hide the message but rather to change its form in a way that would make it appear dignified. While the symbols used in the inscription were scrambled, they were still readable by those who were familiar with the method of substitution used. It means that the inscription was intended for a specific audience who were already familiar with the method rather than as a means of keeping the message secret from all who might view it.

2.1 Substitution cipher

One of the earliest known methods is the simple substitution cipher, which involves replacing each letter of the alphabet with another letter according to a predetermined pattern. There are two types of substitution cipher:

1. Monoalphabetic substitution: a basic cryptography method where each character of the plaintext is replaced with a corresponding character of cipher text. The same substitute symbol or letter is used every time a particular plaintext letter appears. For example, if "A" is substituted with "D," every "A" in the plaintext will be replaced with "D" in the cipher text as shown in **Figure 1**. This makes it vulnerable to frequency analysis attacks as the frequency of each letter in the cipher text will correspond to the frequency of the original letters in the plaintext. Therefore, it is considered a weak encryption method and is no longer used for serious cryptographic applications. However, it can still be used as a simple way to obscure text such as in puzzles or games.

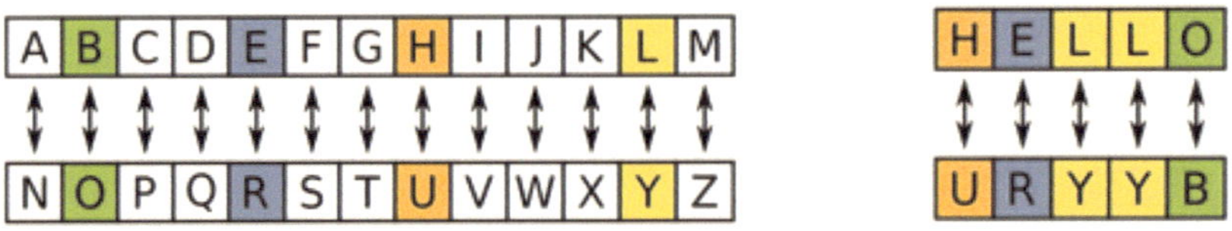

Figure 1.
Monoalphabetic substitution cryptography.

Plain text	ABCDEFGHIJKLMNOPQRSTUVWXYZ
Substitution cipher 1 =	DEFGHIJKLMNOPQRSTUVWXYZABC
Substitution cipher 2 =	GHIJKLMNOPQRSTUVWXYZABCDEF
Substitution cipher 3 =	JKLMNOPQRSTUVWXYZABCDEFGHI
Substitution cipher 4 =	MNOPQRSTUVWXYZABCDEFGHIJKL

Figure 2.
Caesar cipher with 1, 2, 3, and 4 shit to the left.

One of the earliest examples of a monoalphabetic substitution cipher is the Caesar cipher, which was used by Julius Caesar to communicate secretly with his generals. In this cipher, each letter in the plaintext is shifted a certain number of places down the alphabet. For example, if the shift value is three, then the letter A is replaced by D, B is replaced by E, and so on shown in **Figure 2**. The recipient of the message would need to know the shift value to decrypt the message.

Another example of a monoalphabetic substitution cipher is the simple substitution cipher in which each plaintext letter is replaced by a corresponding symbol or letter from a fixed substitution pattern. Unlike the Caesar cipher, the substitution pattern for the simple substitution cipher is not based on a fixed shift value. Instead, the substitution pattern is usually chosen randomly or based on a key provided to the recipient.

Despite being simple to implement, monoalphabetic substitution ciphers are not secure by today's standards as it makes it easier for an attacker to crack the code.

2. Polyalphabetic substitution: It is made up of multiple monoalphabetic substitutions. In this method, a series of monoalphabetic substitutions are performed on the plaintext, using different substitution alphabets for each letter of the plaintext. This helps to make the ciphertext more difficult to crack as the same plaintext letter can be encrypted in different ways depending on its position in the message.

Vigenère cipher is the most known polyalphabetic substitution, which was invented in the sixteenth century and used by the French military for several centuries [14]. The Vigenère cipher uses a series of different alphabets, each generated by shifting the previous alphabet by one letter. The cipher is implemented using the Vigenère square (or table), which is made up of twenty-six distinct cipher alphabets as shown in **Figure 3**. In the header row, the alphabet is written in its normal order. In each subsequent row, the alphabet is shifted one letter to the right until a 26×26 block of letters is formed.

plainText	A	B	C	D	E	F	G	H	I	J	K	L	M	N	O	P	Q	R	S	T	U	V	W	X	Y	Z
1	B	C	D	E	F	G	H	I	J	K	L	M	N	O	P	Q	R	S	T	U	V	W	X	Y	Z	A
2	C	D	E	F	G	H	I	J	K	L	M	N	O	P	Q	R	S	T	U	V	W	X	Y	Z	A	B
3	D	E	F	G	H	I	J	K	L	M	N	O	P	Q	R	S	T	U	V	W	X	Y	Z	A	B	C
4	E	F	G	H	I	J	K	L	M	N	O	P	Q	R	S	T	U	V	W	X	Y	Z	A	B	C	D
5	F	G	H	I	J	K	L	M	N	O	P	Q	R	S	T	U	V	W	X	Y	Z	A	B	C	D	E
6	G	H	I	J	K	L	M	N	O	P	Q	R	S	T	U	V	W	X	Y	Z	A	B	C	D	E	F
7	H	I	J	K	L	M	N	O	P	Q	R	S	T	U	V	W	X	Y	Z	A	B	C	D	E	F	G
8	I	J	K	L	M	N	O	P	Q	R	S	T	U	V	W	X	Y	Z	A	B	C	D	E	F	G	H
9	J	K	L	M	N	O	P	Q	R	S	T	U	V	W	X	Y	Z	A	B	C	D	E	F	G	H	I
10	K	L	M	N	O	P	Q	R	S	T	U	V	W	X	Y	Z	A	B	C	D	E	F	G	H	I	J
11	L	M	N	O	P	Q	R	S	T	U	V	W	X	Y	Z	A	B	C	D	E	F	G	H	I	J	K
12	M	N	O	P	Q	R	S	T	U	V	W	X	Y	Z	A	B	C	D	E	F	G	H	I	J	K	L
13	N	O	P	Q	R	S	T	U	V	W	X	Y	Z	A	B	C	D	E	F	G	H	I	J	K	L	M
14	O	P	Q	R	S	T	U	V	W	X	Y	Z	A	B	C	D	E	F	G	H	I	J	K	L	M	N
15	P	Q	R	S	T	U	V	W	X	Y	Z	A	B	C	D	E	F	G	H	I	J	K	L	M	N	O
16	Q	R	S	T	U	V	W	X	Y	Z	A	B	C	D	E	F	G	H	I	J	K	L	M	N	O	P
17	R	S	T	U	V	W	X	Y	Z	A	B	C	D	E	F	G	H	I	J	K	L	M	N	O	P	Q
18	S	T	U	V	W	X	Y	Z	A	B	C	D	E	F	G	H	I	J	K	L	M	N	O	P	Q	R
19	T	U	V	W	X	Y	Z	A	B	C	D	E	F	G	H	I	J	K	L	M	N	O	P	Q	R	S
20	U	V	W	X	Y	Z	A	B	C	D	E	F	G	H	I	J	K	L	M	N	O	P	Q	R	S	T
21	V	W	X	Y	Z	A	B	C	D	E	F	G	H	I	J	K	L	M	N	O	P	Q	R	S	T	U
22	W	X	Y	Z	A	B	C	D	E	F	G	H	I	J	K	L	M	N	O	P	Q	R	S	T	U	V
23	X	Y	Z	A	B	C	D	E	F	G	H	I	J	K	L	M	N	O	P	Q	R	S	T	U	V	W
24	Y	Z	A	B	C	D	E	F	G	H	I	J	K	L	M	N	O	P	Q	R	S	T	U	V	W	X
25	Z	A	B	C	D	E	F	G	H	I	J	K	L	M	N	O	P	Q	R	S	T	U	V	W	X	Y

Figure 3.
Vigenère square.

Vigenère cipher can be done using the simplest way, which is similar to Caeser cipher or sophisticated way, where keyword is used for the encryption to specify the letter, the keyword is repeated over the length of the plaintext, and each letter of the keyword is used to shift the corresponding letter of the plaintext by a certain number of positions in the alphabet. For example, if you encrypt "security" using the simple way, it will be "TGFYWOAG." But when using the sophisticated way with "IBRI" as a keyword, the cipher text will be "AFTCZJKG." To make the cipher more secure, Vigenère suggested using a different keyword for each message rather than reusing the same keyword over and over again. He also suggested using longer keywords to make the cipher even harder to crack. However, if the length of the keyword is known, it can be easily broken using frequency analysis [15]. **Figure 4** shows an example of onetime pad encryption/decryption.

The onetime pad cipher is not a type of Vigenère cipher. It is a completely different encryption method that is based on using a long, randomly generated key that is at least as long as the plaintext. The key is made up of a series of random symbols, and each symbol is used only once to encrypt one character of the plaintext. Because the key is truly random and used only once, the onetime pad cipher is considered unbreakable, provided that the key is kept secret and destroyed after use by both the sender and the receiver.

The key must be as long as the plaintext for the onetime pad to be unbreakable. Because onetime pad is based on perfect secrecy, which means that the ciphertext provides no information about the plaintext, even if the attacker has unlimited computational power.

Plain text: "LIFE IS SHORT" **Key:** "ENCRYPTIONS"

Plain text	L	I	F	E	I	S	S	H	O	R	T
Plain text value	12	9	6	5	9	19	19	8	15	18	20
One time pad text	E	N	C	R	Y	P	T	I	O	N	S
One time pad value	5	14	3	18	25	16	20	9	15	14	19
Sum of values	17	23	9	23	34	35	39	17	30	32	39
After module subtraction					8	9	13		4	6	13
Cipher text	Q	W	I	W	H	I	M	Q	D	F	M

Cipher text	Q	W	I	W	H	I	M	Q	D	F	M
Cipher text value	17	23	9	23	8	9	13	17	4	6	13
One time pad text	E	N	C	R	Y	P	T	I	O	N	S
One time pad value	5	14	3	18	25	16	20	9	15	14	19
subtracte values	12	9	6	5	-17	-7	-7	8	-11	-8	-6
After module subtraction					9	19	19		15	18	20
Plain text	L	I	F	E	I	S	S	H	O	R	T

Figure 4.
onetime-pad encryption/decryption example.

Generating truly random keys that are as long as the plaintext is a challenging task, and transmitting them securely to the recipient is also a difficult problem. This is why the onetime pad is mostly used in special cases such as diplomatic and intelligence traffic. Also, onetime pad only guarantees confidentiality and not integrity. This means that an attacker who intercepts the ciphertext can not recover the plaintext, but they can easily modify the ciphertext to change the meaning of the message. Onetime pad requires a unique key for every message, and the keys should be securely destroyed after use to prevent reuse.

The Playfair cipher is a polygraphic substitution cipher invented in 1854 by Sir Charles Wheatstone [16]. It was the first cipher that allowed for the encryption of pairs of letters instead of single letters. The Playfair cipher uses a 5 × 5 grid of letters, with each letter of the alphabet appearing once. The letters in the grid are usually chosen using a keyword. The keyword is then written into the grid, and the remaining spaces are filled with the letters of the alphabet in order.

To encrypt plaintext using the Playfair cipher, the plaintext is divided into pairs of letters. If there is an odd number of letters, a dummy letter such as "X" is added at the end. Each pair of letters is then encrypted using the following steps and demonstrated in **Figure 5**:

Step 1: Construct the MATRIX

- Simple way (without keyword)
  - 5 × 5 table
  - Skip letter J
- Sophisticated way (keyword)
  - Keyword has no repeating letter

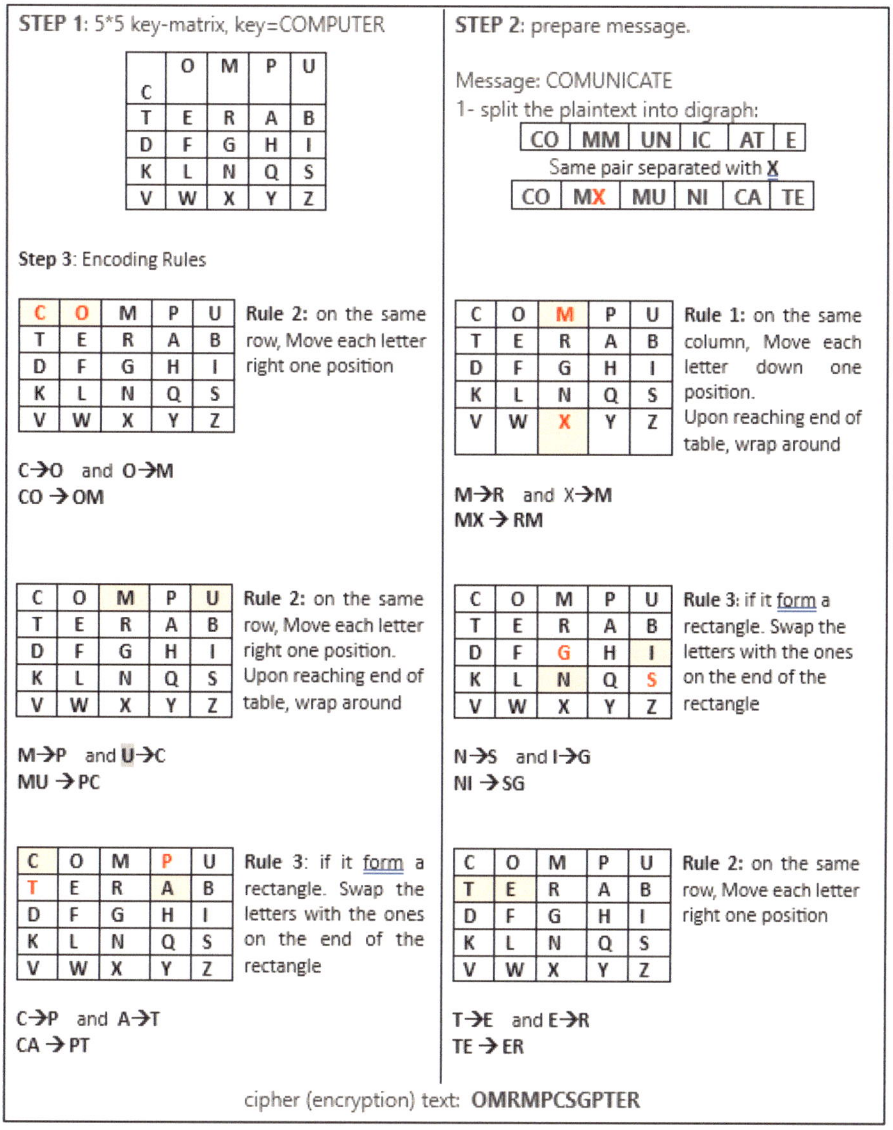

Figure 5.
Playfair cipher steps (A: simple and B: Sophisticated).

- 5 × 5 table
- fill in the remaining letters in alphabetic order (skip letter J)

Step 2: Preparing your message

- Message must be split into pairs
- Repeating plaintext letters that are in the same pair are separated with X
- If there is an odd letter at the end of the message insert the letter X

Step 3: Encoding Rules:

- Rule 1: If they fall in the same **column**
    - Move each letter down one position
    - Upon reaching end of table, wrap around
- Rule 2: If they fall in the same **row**
    - Move each letter right one position
    - Upon reaching end of table, wrap around
- Rule 3: If it forms a **rectangle**
    - Swap the letters with the ones on the end of the rectangle

An electromechanical machine developed in 2017 [17] that used a rotating disc with an embedded key to encode a substitution table that changed with every new character typed. This device was the first example of a rotor machine. The following year, a German engineer, invented the Enigma machine [18], which used multiple rotors instead of one. Initially designed for commercial use, the German military soon recognized the potential of the Enigma machine and began using it to send coded transmissions.

2.2 Transposition cipher

Transposition cipher is an earlier method, where the letters of the message are rearranged according to a certain pattern, but the letters themselves are not changed as shown in **Figure 6**. Unlike substitution ciphers, which replace plaintext characters with different symbols or letters, transposition ciphers do not change the characters themselves. Instead, they simply reorder the characters to create a new message. The security of a transposition cipher is based on the difficulty of reconstructing the original message from the reordered characters without knowledge of the used transposition algorithm.

The Rail Fence cipher is a type of transposition cipher that was first used during the American Civil War. The technique involves writing the plaintext diagonally on a grid, then reading the letters in a zigzag pattern along the rows of the grid to produce the ciphertext. The number of rows in the grid can be adjusted to increase the complexity of the cipher.

Plaintext	A	L	G	O	R	I	T	H	M	S
Position	1	2	3	4	5	1	2	3	4	5
Key	3	5	2	1	4	3	5	2	1	4
Ciphertext	G	R	L	A	O	H	S	T	I	M

Figure 6.
Transposition cipher example.

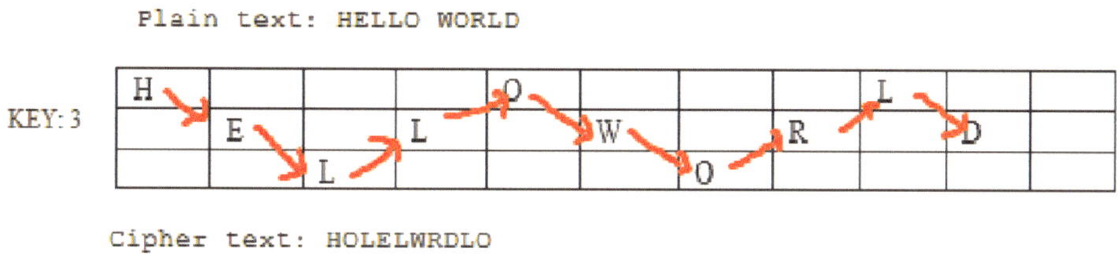

Figure 7.
Rail Fence encryption example.

For example, suppose we want to encrypt the message "HELLO WORLD" using a Rail Fence cipher with three rows. Write the letters on a grid as shown in **Figure** 7.

To decrypt the message, we would write the ciphertext diagonally on a grid, then read the letters in the same zigzag pattern along the rows of the grid to recover the plaintext.

While these ancient methods of cryptography may seem primitive by today's standards, they laid the foundation for the development of more complex encryption techniques in the future. The principles of substitution and transposition ciphers are still used in modern cryptography, and the need for secure communication continues to drive the evolution of cryptographic algorithms.

3. Symmetric key cryptography

Symmetric key cryptography schemes are categorized as stream ciphers or block ciphers. Stream ciphers work on a single bit at a time and execute some form of feedback structure so that the key is repeatedly changing. A block cipher encrypts one block at a time utilizing the same key on each block. In general, the same plaintext block will continually encrypt to the same ciphertext when using the similar key in a block cipher, whereas the same plaintext will encrypt to different ciphertext.

The history of symmetric key cryptography can be traced back to the days of Julius Caesar, who used a simple substitution cipher to protect his military communications. Over time, various types of symmetric key encryption algorithms were developed, such as the Vigenère cipher, which used a polyalphabetic substitution method, and the Enigma machine, which used a combination of substitution and transposition methods.

3.1 Data encryption standard (DES)

One of the most widely used symmetric key encryption algorithms is the data encryption standard (DES), which was developed by IBM in the 1970s and adopted in 1977 by the National Bureau of Standards, now the National Institute of Standards and Technology (NIST), as Federal Information Processing Standard. U.S. Data are encrypted in DES using a block cipher method to encrypt data in 64-bit blocks, with a 56-bit key. The algorithm transforms 64-bit input (plaintext) in a series of steps into a 64-bit output (ciphertext). The same steps, with the same key, are used to reverse the encryption. The encryption process in DES involves three phases:

1. Initial permutation (IP): The 64-bit input plaintext is shuffled (rearranged) according to a fixed permutation table to produce the permuted input. The initial

Initial Permutation (IP)								Inverse Initial Permutation (IP^{-1})							
58	50	42	34	26	18	10	2	40	8	48	16	56	24	64	32
60	52	44	36	28	20	12	4	39	7	47	15	55	23	63	31
62	54	46	38	30	22	14	6	38	6	46	14	54	22	62	30
64	56	48	40	32	24	16	8	37	5	45	13	53	21	61	29
57	49	41	33	25	17	9	1	36	4	44	12	52	20	60	28
59	51	43	35	27	19	11	3	35	3	43	11	51	19	59	27
61	53	45	37	29	21	13	5	34	2	42	10	50	18	58	26
63	55	47	39	31	23	15	7	33	1	41	9	49	17	57	25

Figure 8.
The initial permutation and its inverse.

permutation and its inverse are defined by tables that indicate the position of each bit in the input to the output as shown in **Figure 8**. The permutation tables are used in the encryption and decryption processes to rearrange the bits of the input according to the specified permutation.

2. 16 rounds of a complex key-dependent round function: Each round involves a substitution and a permutation. The 56-bit key is used to generate 16 round subkeys, each consisting of 48 bits, which are used in the round function. The output of the 16th round consists of 64 bits that are a function of the input plaintext and the key. The left and right halves of the output are swapped to produce the pre-output. **Figure 9** shows the internal structure of a single round, focusing on the left-hand side of the diagram. The main steps are:

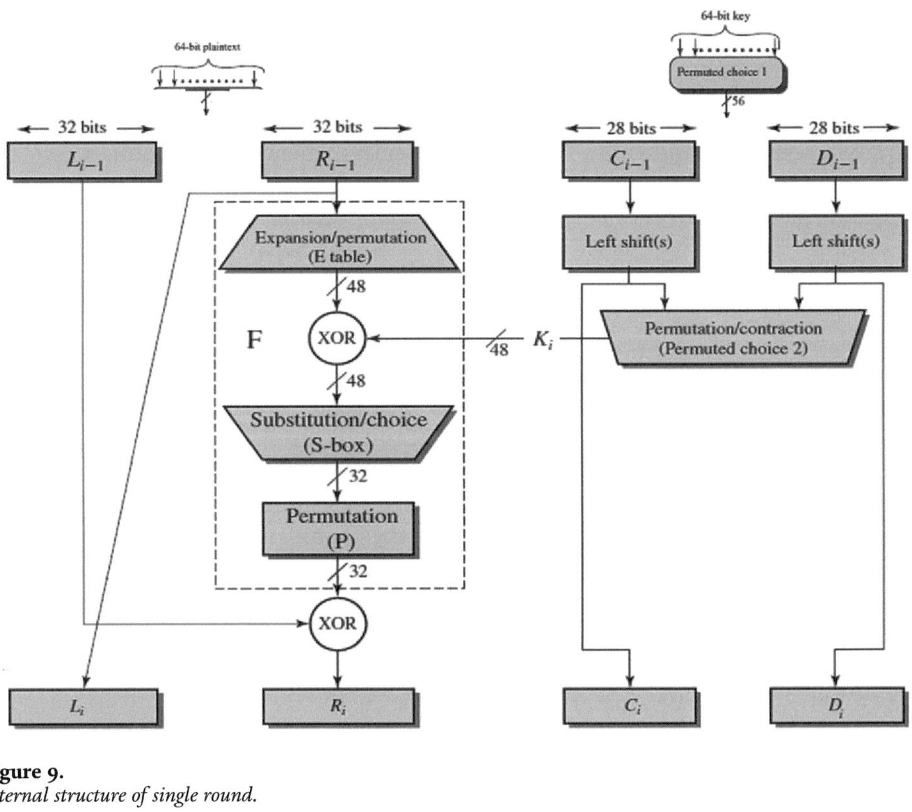

Figure 9.
Internal structure of single round.

32	1	2	3	4	5
4	5	6	7	8	9
8	9	10	11	12	13
12	13	14	15	16	17
16	17	18	19	20	21
20	21	22	23	24	25
24	25	26	27	28	29
28	29	30	31	32	1

Figure 10.
Expansion permutation table.

A. Separation: The left and right halves of each 64-bit intermediate value are treated as separate 32-bit quantities, labeled L (left) and R (right).

B. Expansion: The input key for each round is 48 bits and the right side (R) is 32 bits. In order to XOR Ki with Ri, we need to expand the length of Ri to 48 bits. The expansion table in **Figure 10** is used for this purpose.

C. Key mixing: The 48-bit result from the expansion step is XORed with a 48-bit subkey derived from the main 56-bit key. The subkey is generated using a combination of permutation and substitution operations on the main key. As shown in **Figure 9**, the subkey generation in DES involves the following steps:

- The 64-bit key is permuted using a fixed permutation called the permutation choice 1 (PC-1) as shown in **Figure 11**. The output of this step is a 56-bit key, where eight of the bits are parity bits and are not used in the encryption process.

Permuted Choice One(PC-1)

57	49	41	33	25	17	9
1	58	50	42	34	26	18
10	2	59	51	43	35	27
19	22	3	60	52	44	36
63	55	47	39	31	23	15
7	62	54	46	38	30	22
14	6	61	53	45	37	29
21	13	5	28	20	12	4

Permuted Choice Two(PC-2)

14	17	11	24	1	5	3	28
15	6	21	10	23	19	12	4
26	8	16	7	27	20	13	2
42	52	31	37	47	55	30	40
51	45	33	48	44	49	39	56
34	53	46	42	50	36	29	32

Figure 11.
Tables used in subkeys generation.

- The 56-bit key is then split into two 28-bit halves, C0 and D0.
- Each of the halves is subjected to a series of circular shifts or rotations. In particular, for rounds 1, 2, 9, and 16, the shifts are one bit, while for all other rounds, the shifts are two bits.
- After each shift, the two halves are combined to form a 56-bit value, which is then permuted using a fixed permutation called the permutation choice 2 (PC-2) as shown in **Figure 9**. The output of this step is a 48-bit subkey.
- This process is repeated for each round of the encryption process, resulting in a total of 16 subkeys.
- The subkeys are used in the encryption process as inputs to the round function, which combines them with the plaintext to produce the ciphertext.

D. Substitution: This 48-bit result passes through a substitution function that produces a 32-bit output. The S-boxes, also known as substitution boxes, are the only nonlinear elements in the DES design. The S-boxes are used to introduce confusion in the ciphertext by replacing each block of 6 bits of the input with a different 4-bit output. There are 8S-boxes in DES as shown in **Figure 12**, each taking a 6-bit input and producing a 4-bit output. Each row of an S-box defines a substitution for a specific 4-bit input value, while the column of the S-box defines the output value for that input value based

S_1	14	4	13	1	2	15	11	8	3	10	6	12	5	9	0	7
	0	15	7	4	14	2	13	1	10	6	12	11	9	5	3	8
	4	1	14	8	13	6	2	11	15	12	9	7	3	10	5	0
	15	12	8	2	4	9	1	7	5	11	3	14	10	0	6	13
S_2	15	1	8	14	6	11	3	4	9	7	2	13	12	0	5	10
	3	13	4	7	15	2	8	14	12	0	1	10	6	9	11	5
	0	14	7	11	10	4	13	1	5	8	12	6	9	3	2	15
	13	8	10	1	3	15	4	2	11	6	7	12	0	5	14	9
S_3	10	0	9	14	6	3	15	5	1	13	12	7	11	4	2	8
	13	7	0	9	3	4	6	10	2	8	5	14	12	11	15	1
	13	6	4	9	8	15	3	0	11	1	2	12	5	10	14	7
	1	10	13	0	6	9	8	7	4	15	14	3	11	5	2	12
S_4	7	13	14	3	0	6	9	10	1	2	8	5	11	12	4	15
	13	8	11	5	6	15	0	3	4	7	2	12	1	10	14	9
	10	6	9	0	12	11	7	13	15	1	3	14	5	2	8	4
	3	15	0	6	10	1	13	8	9	4	5	11	12	7	2	14
S_5	2	12	4	1	7	10	11	6	8	5	3	15	13	0	14	9
	14	11	2	12	4	7	13	1	5	0	15	10	3	9	8	6
	4	2	1	11	10	13	7	8	15	9	12	5	6	3	0	14
	11	8	12	7	1	14	2	13	6	15	0	9	10	4	5	3
S_6	12	1	10	15	9	2	6	8	0	13	3	4	14	7	5	11
	10	15	4	2	7	12	9	5	6	1	13	14	0	11	3	8
	9	14	15	5	2	8	12	3	7	0	4	10	1	13	11	6
	4	3	2	12	9	5	15	10	11	14	1	7	6	0	8	13
S_7	4	11	2	14	15	0	8	13	3	12	9	7	5	10	6	1
	13	0	11	7	4	9	1	10	14	3	5	12	2	15	8	6
	1	4	11	13	12	3	7	14	10	15	6	8	0	5	9	2
	6	11	13	8	1	4	10	7	9	5	0	15	14	2	3	12
S_8	13	2	8	4	6	15	11	1	10	9	3	14	5	0	12	7
	1	15	13	8	10	3	7	4	12	5	6	11	0	14	9	2
	7	11	4	1	9	12	14	2	0	6	10	13	15	3	5	8
	2	1	14	7	4	10	8	13	15	12	9	0	3	5	6	11

Figure 12.
S-boxes used in the substitution step in DES.

on the remaining 2 bits of the input. This allows for a total of 16 x 4 = 64 possible substitutions in each S-box.

E. Permutation: The 32-bit outputs from the S-boxes are then concatenated and subjected to a fixed permutation using the P-box permutation.

3. Final permutation (IP-1): The pre-output is shuffled according to another fixed permutation table, which is the inverse of the initial permutation, to produce the 64-bit cipher text. The figure shows the internal structure of a single round.

The main steps summarized in **Figure 13**. The DES key generates 48 bits long 16 round keys from the initial 56 bit key. These keys are used in each round of the encryption process to modify the plaintext. The key involves applying a series of operations, including a permutation, a compression function, and left shifts, to the 56-bit key. The resulting subkeys are used one at a time in each round of the encryption process.

However, due to its small key size, DES is now considered insecure [19] and has been replaced by the advanced encryption s (AES).

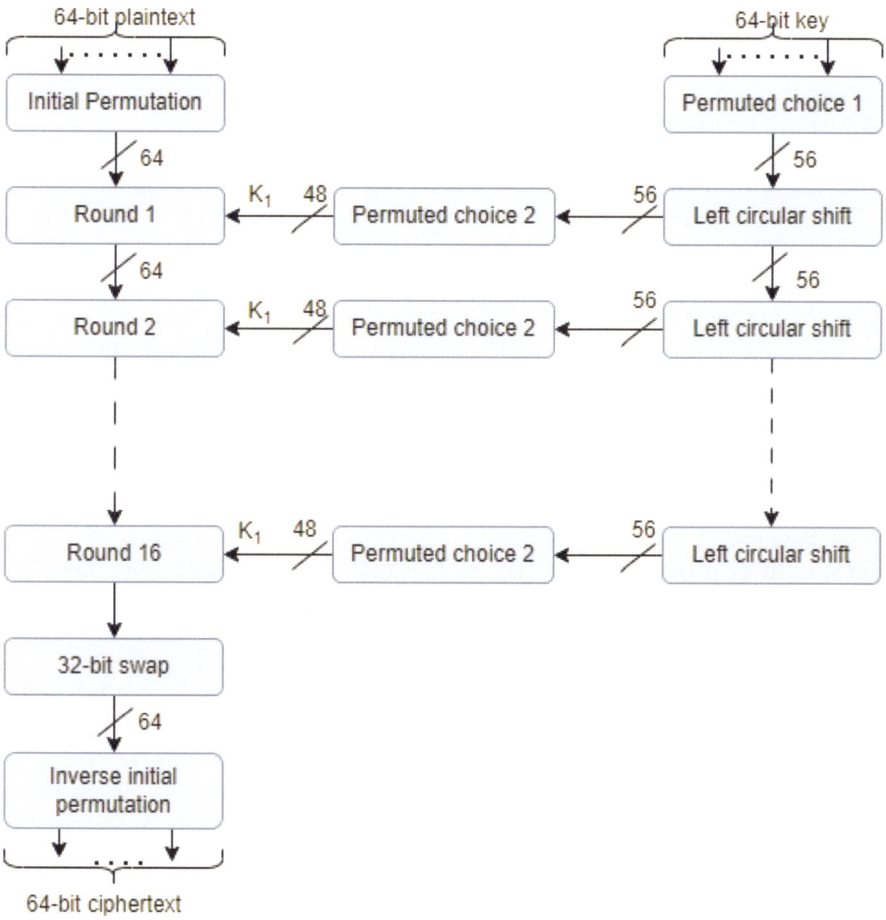

Figure 13.
DES Algorithm steps.

To be more precise, 3DES (also known as Triple DES or TDEA) is a symmetric key cipher that uses the DES algorithm three times in succession to increase its security [1, 20]. The standard 3DES encryption process can be described as follows:

1. The plaintext is encrypted using the first 56-bit key (K1) with the DES algorithm to produce a ciphertext.

2. The ciphertext from step 1 is decrypted using the second 56-bit key (K2) with the DES algorithm to produce an intermediate value.

3. The intermediate value from step 2 is encrypted again using the third 56-bit key (K3) with the DES algorithm to produce the final ciphertext.

Thus, 3DES involves encrypting the plaintext with K1, decrypting the result with K2, and encrypting again with K3. The three keys K1, K2, and K3 are usually independent keys generated randomly, although some variants of 3DES use a "keying option" that allows for fewer keys to be used while still maintaining a higher level of security.

While 3DES is slower than DES due to its triple encryption process, it is still considered a relatively fast algorithm and can be implemented in hardware, as well as software. Also, due to its small key size, DES is now considered insecure [19] and has been replaced by the advanced encryption standard (AES).

3.2 Advanced encryption standard (AES)

The AES (Advanced Encryption Standard) is a symmetric block cipher that operates on fixed-size 128-bit blocks and supports key sizes of 128, 192, and 256 bits. It was standardized by NIST (National Institute of Standards and Technology) in 2001 as a replacement for the aging DES (Data Encryption Standard) cipher.

The AES was selected from a pool of 15 candidate algorithms that were submitted in response to a call for proposals issued by NIST in 1997 [21]. The selection process involved several rounds of analysis and testing, culminating in the selection of Rijndael [22], a cipher developed by Belgian cryptographers Joan Daemen and Vincent Rijmen, as the winner.

The AES encryption and decryption algorithms use a series of rounds, where all operations are performed on 8-bit bytes (one Word) (**Figure 14**). Each round of processing works on the input state array and produces an output state array. The output state array produced by the last round is rearranged into a 128-bit output block. The state array is a 4×4 matrix of bytes that represents the input block. Each round, the state array is modified by a series of operations that include byte substitution, permutation, and arithmetic operations over a finite field as shown in the figure below. After the final round, the state array contains the encrypted or decrypted data, which are then copied to an output matrix to produce the final ciphertext or plaintext block.

Each round can be described in four functions as shown in **Figure 14**. These four functions are combined in a specific order to form a round, and multiple rounds are performed on the input block to produce the final ciphertext block. The number of rounds depends on the key size and block size, with 10, 12, or 14 rounds used for 128-bit, 192-bit, and 256-bit keys, respectively. The functions are:

1. **SubBytes**: The substitute bytes stage of AES uses a fixed S-box, which is a 256-byte lookup table, to perform a byte-by-byte substitution of the input block. The

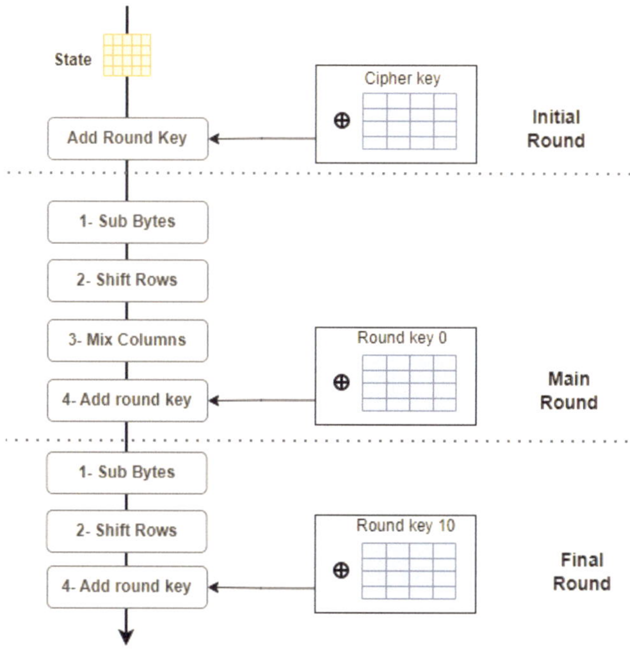

Figure 14.
The structure of AES algorithm.

S-box is designed so that each input byte is replaced by a unique output byte. The inverse S-box is used in the decryption process, which maps each output byte back to its original input byte. The S-box is a nonlinear component of the AES algorithm, which helps to increase the resistance of the cipher to various attacks. For example, 19 will be mapped to the value crossed between row 1 and column 9, which is equal to D4 in the S-Box as shown in **Figure 15**.

X \ Y	0	1	2	3	4	5	6	7	8	9	A	B	C	D	E	F
0	63	7c	77	7b	f2	6b	6f	c5	30	1	67	2b	fe	d7	ab	76
1	ca	82	c9	7d	fa	59	47	f0	ad	d4	a2	af	9c	a4	72	c0
2	b7	fd	93	26	36	3f	f7	cc	34	a5	e5	f1	71	d8	31	15
3	4	c7	23	c3	18	96	5	9a	7	12	80	e2	eb	27	b2	75
4	9	83	2c	1a	1b	6e	5a	a0	52	3b	d6	b3	29	e3	2f	84
5	53	d1	0	ed	20	fc	b1	5b	6a	cb	be	39	4a	4c	58	cf
6	d0	ef	aa	fb	43	4d	33	85	45	f9	2	7f	50	3c	9f	a8
7	51	a3	40	8f	92	9d	38	f5	bc	b6	da	21	10	ff	f3	d2
8	cd	0c	13	ec	5f	97	44	17	c4	a7	7e	3d	64	5d	19	73
9	60	81	4f	dc	22	2a	90	88	46	ee	b8	14	de	5e	0b	db
A	e0	32	3a	0a	49	6	24	5c	c2	d3	ac	62	91	95	e4	79
B	e7	c8	37	6d	8d	d5	4e	a9	6c	56	f4	ea	65	7a	ae	8
C	ba	78	25	2e	1c	a6	b4	c6	e8	dd	74	1f	4b	bd	8b	8a
D	70	3e	b5	66	48	3	f6	0e	61	35	57	b9	86	c1	1d	9e
E	e1	f8	98	11	69	d9	8e	94	9b	1e	87	e9	ce	55	28	df
F	8c	a1	89	0d	bf	e6	42	68	41	99	2d	0f	b0	54	bb	16

Figure 15.
S-Box used in AES.

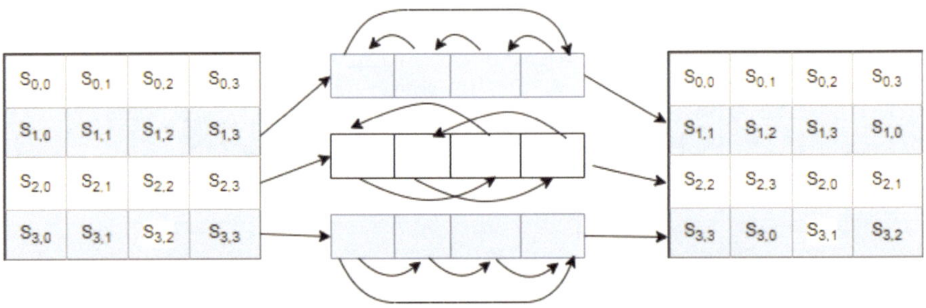

Figure 16.
ShiftRows operation and its output (with example).

2. **ShiftRows**: The shiftRows stage is a permutation step that cyclically shifts the bytes in each row of the state array by a certain number of bytes. This operation is applied to each row independently, with no mixing of the bytes between the rows. The number of bytes shifted is determined by the row number: the first row is not shifted at all, the second row is shifted by one byte to the left, the third row is shifted by two bytes to the left, and the fourth row is shifted by three bytes to the left as shown in **Figure 16**.

   This operation provides diffusion of the input data, which increases the security of the cipher. The inverse operation, used for decryption, is a cyclic shift to the right instead of the left so that the original byte positions are restored.

3. **MixColumns**: each column of the state array is treated as a polynomial over the finite field GF(2^8), where each byte is a coefficient of the polynomial. The bytes are then multiplied by a fixed polynomial, and the result is reduced modulo another fixed polynomial. This transformation ensures that each byte in a column is dependent on all four bytes in the same column as demonstrated in **Figure 17**.

   The multiplication and reduction are done using a pre-computed table of values. The table is constructed in such a way that multiplication is reduced to a simple table lookup and XOR operation.

   During decryption, the inverse operation of MixColumns is performed. This involves multiplying each column by a different fixed polynomial and reducing the result modulo another fixed polynomial.

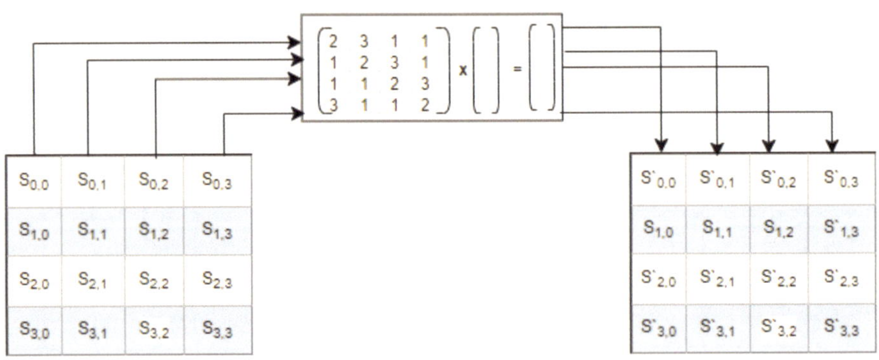

Figure 17.
Mix column function.

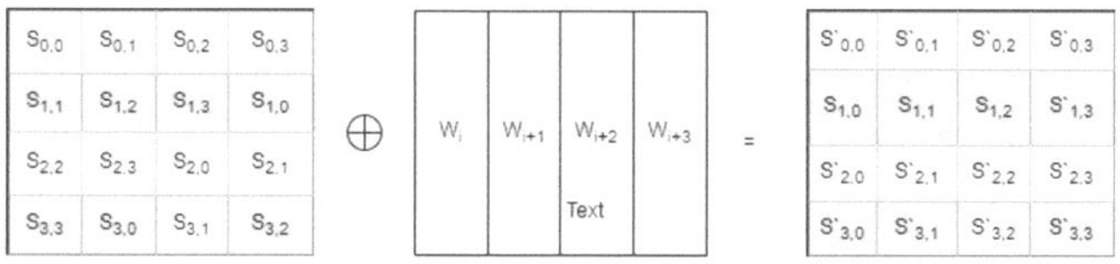

Figure 18.
Description of the AddRoundkey in AES.

4. **AddRoundkey**: Each byte of the current block is XORed with the corresponding byte of the round key. The round key is derived from the main encryption key using a key schedule algorithm, which generates a set of round keys for each round of encryption. This stage serves to add a layer of confusion to the encryption process, making it more difficult to analyze and break the cipher. **Figure 18** describe the AddRoundkey process in AES.

The AES key expansion algorithm takes as input a 128-bit (16-byte) key and generates a sequence of round keys, one for each round of the AES encryption process. The key expansion algorithm uses a key schedule to generate these round keys, which involves performing a series of operations on the input key to generate an expanded key.

The key schedule begins by copying the input key into the first four words of the key schedule. Then, the key expansion algorithm applies a series of operations to the last four words of the current key schedule to generate the next four words. This process is repeated until the key schedule contains the necessary number of round keys for the specified key size. For example, for a 128-bit key, the key schedule will generate 11 round keys, one for each of the 10 rounds of AES encryption plus an initial round key. For a 192-bit key, the key schedule will generate 13 round keys, and for a 256-bit key, the key schedule will generate 15 round keys.

In the key expansion algorithm, the first word in each group of four undergoes a series of operations before being XORed with the word from fourth positions back. These operations include a one-byte circular left shift (RotWord), byte substitution using the S-box (SubWord), and XORing with a round constant (Rcon[j]), the values of Rcon[j] shown in **Figure 19**. In the 256-bit key/14-round version, an additional step is performed on the middle word. The steps are:

1. RotWord performs a one-byte circular left shift on a word.

2. SubWord performs a byte substitution on each byte of its input word, using the S-box.

3. The result of steps 1 and 2 is XORed with a round constant, Rcon[j].

The AES cipher is widely used in various applications, including secure communications, data storage, and authentication. Its security has been extensively analyzed, and it is considered to be highly secure against various types of attacks.

J	1	2	3	4	5	6	7	8	9	10
Rcon[j]	01	02	04	08	10	20	40	80	1B	36

Figure 19.
The values of Rcon[j] in hexadecimal.

3.3 More symmetric algorithms

- Blowfish [23]: A symmetric key block cipher that uses variable-length keys (up to 448 bits) and a block size of 64 bits. Blowfish is widely used in cryptographic applications and is known for its fast encryption and decryption speed.
- Twofish [24]: A symmetric key block cipher that is a successor to Blowfish. It uses a block size of 128 bits and supports key sizes up to 256 bits. Twofish is considered a strong and secure encryption algorithm but is slower than some other algorithms.
- Rivest Cipher 4 (RC4) [25]: A symmetric key stream cipher that is widely used in wireless networks, secure socket layer (SSL), and other applications. RC4 uses a variable-length key (up to 2048 bits) to generate a stream of pseudo-random bytes, which are XORed with the plaintext to produce the ciphertext. However, RC4 has been found to be vulnerable to attacks and is now considered insecure for many applications.

3.4 Mode of operation

Since block ciphers operate on fixed-size blocks of data, they cannot be directly used to encrypt or decrypt messages that are larger than the block size. A mode of operation is a technique used to apply a block cipher to encrypt or decrypt data that is larger than the block size of the cipher.

Modes of operation are used to overcome this limitation by allowing the encryption or decryption of data that is larger than the block size of the cipher. These modes provide methods to break up the input message into blocks, and then apply the block cipher to each block. This process is typically performed using feedback mechanisms that generate input for each subsequent block, based on the output of the previous block.

There are several modes of operation defined by NIST, each with its own strengths and weaknesses and suitable for different types of applications. For example, some modes are designed to provide confidentiality, while others also provide message integrity and authentication. The five modes of operation defined by NIST are:

1. Electronic codebook (ECB): This is the simplest mode of operation, where each block of plaintext is encrypted independently with the same key as shown in **Figure 20**. However, it is not suitable for encrypting large amounts of data or data with a predictable structure. It suffers from the lack of diffusion, which means that identical plaintext blocks will result in identical ciphertext blocks.

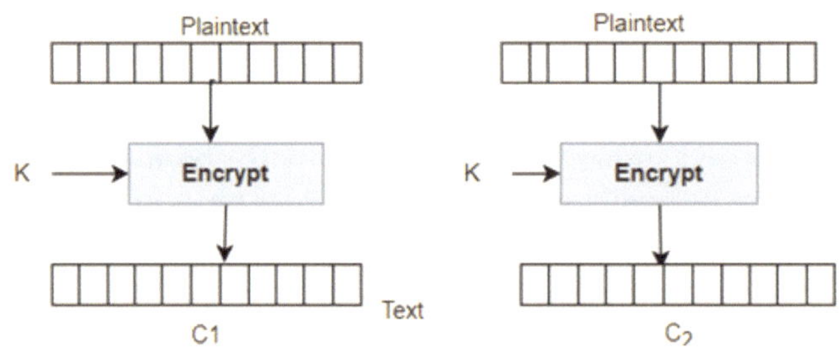

Figure 20.
ECB mode encryption.

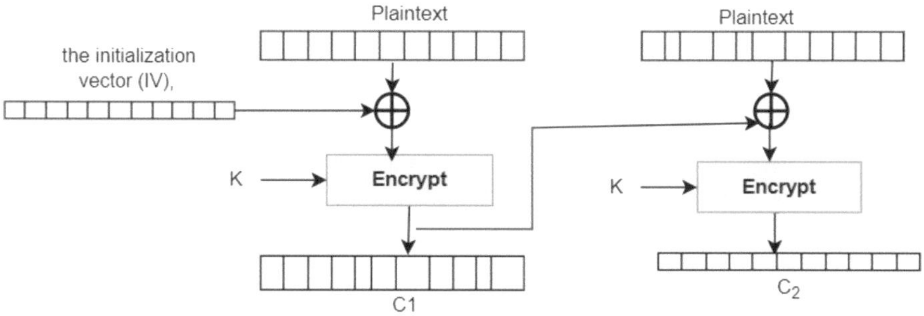

Figure 21.
CBC mode encryption.

This makes it vulnerable to attacks as patterns in the plaintext can be easily observed in the ciphertext. For example, an image encrypted with ECB mode will have visible patterns and blocks, making it easy for an attacker to identify certain parts of the image even without decrypting it. Therefore, it is not recommended to use ECB mode for encrypting lengthy messages or sensitive data.

2. Cipher block chaining (CBC): The cipher block chaining (CBC) mode of operation addresses the issue of repetitive plaintext blocks in ECB mode. This mode XORs each plaintext block with the previous ciphertext block before encryption as shown in **Figure 21**. This helps to provide diffusion and makes the encryption process more secure than ECB. Itis worth noting that the sequential nature of CBC encryption can also be an advantage in some cases as it provides a natural form of authentication. If a ciphertext block is corrupted or modified during transmission, the corresponding plaintext block will be affected, and the error will propagate through the rest of the decryption process, making it easier to detect tampering.

   However, one-bit change in a plaintext or IV affects all following ciphertext blocks can also be a weakness. This can make it difficult to implement certain types of secure communications protocols such as those that require random access to encrypted data. Additionally, CBC requires a secure and unpredictable initialization vector (IV) for each message, which can be challenging to generate and transmit securely in some scenarios. Finally, as with any mode of operation that relies on a shared secret key, CBC is vulnerable to attacks that exploit weaknesses in the underlying block cipher or key management protocols.

3. Cipher feedback (CFB): In this mode, the block cipher is used as a feedback mechanism to create a stream cipher. The plaintext is XORed with the output of the block cipher, and the result is encrypted to produce the ciphertext as shown in **Figure 22**. This mode allows for variable-length plaintext and provides a self-synchronizing stream cipher. The initial value is called the initialization vector (IV), and it is used to seed the process. The size of the shift registers determines the amount of feedback. For example, if $s = 8$, the encryption process operates on an 8-bit subset of the plaintext block at a time. If $s = n$, then the entire plaintext block is used at once.

   One advantage of CFB mode is that it allows for error propagation to be contained. If a bit error occurs during transmission, only the block that contains the error is affected. The other blocks remain unchanged. However, one disadvantage of CFB mode is that it is sequential, which means that it cannot be parallelized.

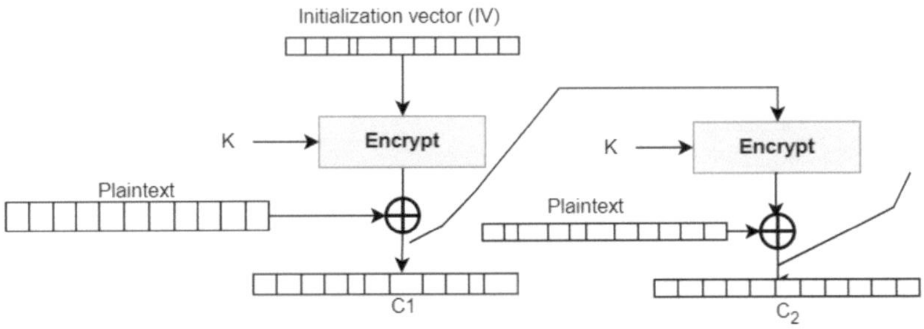

Figure 22.
CFB mode encryption.

4. Output feedback (OFB): OFB mode operates on full blocks of plaintext and ciphertext such as other block cipher modes of operation. However, instead of encrypting the plaintext, the block cipher is used to encrypt an IV to generate a keystream. The keystream is then XORed with the plaintext to produce the ciphertext. The key stream is generated independently for each block, so the encryption and decryption can be parallelized as shown in **Figure 23**. The main difference between OFB and CFB is that OFB generates a key stream that is independent of the plaintext, while CFB uses the ciphertext as feedback to generate the key stream.

5. Counter (CTR): This mode encrypts a counter value with a block cipher to produce a keystream, which is then XORed with the plaintext to produce the ciphertext. This mode is similar to OFB, but it allows for parallel encryption and decryption and can be used for random. The counter is incremented for each block of plaintext, and the resulting keystream is used to encrypt that block, see **Figure 24**. The advantage of the CTR mode is that it allows for parallel encryption and decryption of blocks since the keystream is generated independently of the plaintext or ciphertext. This can lead to significant speed improvements over other modes, particularly for large messages.

   One potential drawback of CTR mode is the need to ensure that the counter values are never repeated as this could compromise the security of the encryption. This can be achieved by using a unique counter value for each block of plaintext, for example by using a nonce (a number used only once) as part of the counter value.

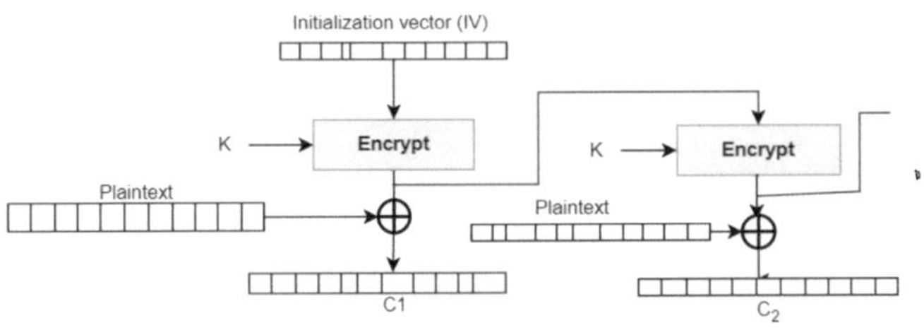

Figure 23.
OFB mode encryption.

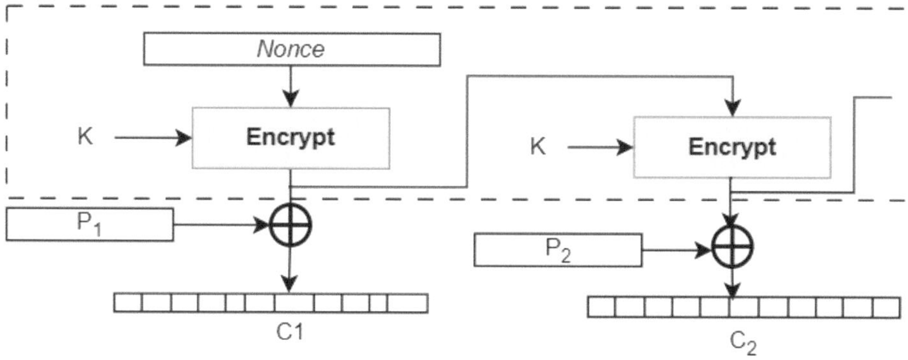

Figure 24.
Counter mode encryption.

4. Asymmetric key cryptography

Asymmetric key cryptography, also known as public-key cryptography, is a cryptographic system that uses a pair of keys to encrypt and decrypt data. The pair of keys consists of a public key, which is known to everyone, and a private key, which is kept secret by its owner. The public key is used for encrypting the data, while the private key is used for decrypting the data. Unlike symmetric key cryptography, where the same key is used for both encryption and decryption, in asymmetric key cryptography, the two keys are mathematically related, but it is computationally infeasible to derive the private key from the public key.

The main advantage of asymmetric key cryptography is that it provides a secure method of communication between two parties without the need for a pre-shared secret key. Asymmetric key cryptography is used in many applications, including digital signatures, key exchange, and encryption of sensitive data.

Some examples of asymmetric key cryptographic algorithms include RSA [26], Diffie-Hellman [27], and elliptic curve cryptography (ECC) [28]. These algorithms are widely used in various applications, including secure communication, digital signatures, and online transactions [29].

4.1 RSA

RSA is a widely used public-key cryptosystem. It is been named after its inventors Ron Rivest, Adi Shamir, and Leonard Adleman. Its security is based on the difficulty of factoring large integers, which serves as the foundation for its mathematical operation. RSA has been used for over four decades and is still considered a secure and practical public-key cryptosystem. RSA involves the generation of a public and a private key pair. The public key is distributed to others, while the private key is kept secret. The public key can be used to encrypt messages that only the owner of the private key can decrypt.

The security of RSA is based on the fact that factoring large integers is a difficult problem, and the larger the key size, the more difficult it becomes. RSA keys typically range in size from 1024 to 4096 bits. We can say that RSA is widely accepted and implemented in various applications such as secure communication, digital signatures, and key exchange [30]. RSA encryption and decryption are performed as follows:

- **Key generation**:

  1. Choose two large prime numbers p and q.
  2. Calculate n = p * q and φ(n) = (p–1) * (q–1).
  3. Choose an integer e such that 1 < e < φ(n) and gcd(e, φ(n)) = 1. This value is called the public exponent.
  4. Compute d, the multiplicative inverse of e modulo φ(n). This value is called the private exponent.

- **Encryption**:

  1. Represent the plaintext M as a positive integer less than n.
  2. Compute the ciphertext C as C = Me mod n.

- **Decryption**: Compute the plaintext M as M = Cd mod n.

The security of RSA is based on the difficulty of factoring large composite numbers into their prime factors. Breaking RSA encryption requires factoring the modulus n into its two prime factors p and q, which is a computationally intensive task for large values of n. Therefore, the security of RSA increases as the size of the keys and the modulus increase.

4.2 Diffie-Hellman

Diffie-Hellman (DH) is a key exchange algorithm that allows two parties to establish a shared secret key over an insecure channel. It was developed by Whitfield Diffie and Martin Hellman in 1976 and is based on the discrete logarithm problem in modular arithmetic.

In DH, each party generates a public-private key pair. The public keys are exchanged and used to derive a shared secret key. The derivation of the key involves modular exponentiation and is based on the fact that the discrete logarithm problem is believed to be hard. The DH protocol works as follows:

1. Alice and Bob publicly agree on a large prime number p and a primitive root of p, denoted by g.
2. Alice randomly chooses a secret integer a and calculates A = g^a mod p. She sends A to Bob.
3. Bob randomly chooses a secret integer b and calculates B = g^b mod p. He sends B to Alice.
4. Alice computes the shared secret key as K = B^a mod p.
5. Bob computes the shared secret key as K = A^b mod p.

6. Alice and Bob now have a shared secret key that can be used for symmetric encryption.

The security here relies on the fact that computing the discrete logarithm of g mod p is computationally infeasible. This means that an attacker who intercepts A and B cannot calculate a or b, and therefore cannot compute the shared secret key K.

The DH algorithm can be used for secure communication by combining it with a symmetric encryption algorithm. The shared secret key derived using DH is used as the key for the symmetric encryption algorithm, providing confidentiality for communication. Widely used in many cryptographic protocols such as Secure Socket Layer (SSL)/Transport Layer Security (TLS), Secure Shell Protocol (SSH), and Virtual private networks (VPNs) [31, 32]. However, it does not provide authentication [32], and therefore a man-in-the-middle attack is possible if the channel is not authenticated. To address this issue, DH is often used in combination with digital signatures or other authentication mechanisms [33].

5. Hash functions

A hash function is a one-way function that takes an input (also known as the message or data) of arbitrary length and produces a fixed-size output, typically represented as a sequence of bytes. The output is often referred to as the hash or message digest. A good hash function should have the following properties:

- Deterministic: The same input should always produce the same output.
- Uniform: The output should appear to be random and uniformly distributed, even if the input has patterns or biases.
- One-way: It should be computationally infeasible to derive the input data from the hash value.
- Collision-resistant: It should be computationally infeasible to find two different input values that produce the same hash output.

Hash functions are commonly used in various security applications such as password storage, digital signatures, and message authentication codes.

6. Digital signatures

Digital signatures are used to ensure the authenticity, integrity, and non-repudiation of a digital document or message. The process of creating a digital signature involves applying a mathematical algorithm to the message or document using the signer's private key. The resulting value, known as the signature, is unique to both the message and the signer's private key.

The receiver of the message or document can verify the signature using the signer's public key, which confirms that the message was indeed sent by the signer and that it has not been altered since it was signed.

Digital signatures can be used in a variety of applications, including software updates, online transactions, and legal documents. They provide a means of verifying the identity of the sender, ensuring the integrity of the message or document, and preventing the sender from denying that they sent the message or document.

7. Future of cryptography

Cryptography has come a long way since its early beginnings, and it continues to play a critical role in securing our digital world today. The advancement of technology has led to more complex and sophisticated encryption methods, which have become essential for protecting sensitive information such as financial transactions, personal data, and confidential communication. With the rise of the internet and mobile technology, cryptography has become more important than ever. It is used in everything from e-commerce to social media to secure online communication [34]. As technology continues to evolve, so will the field of cryptography, and new techniques and algorithms will be developed to stay ahead of emerging threats. The future of cryptography holds great promise as researchers work to develop quantum-resistant encryption and new methods for securing blockchain technology. As we rely more and more on digital communication and storage, the role of cryptography in securing our data will only become more critical.

7.1 Quantum cryptography

Quantum computers have the potential to break many of the current cryptographic schemes that rely on the difficulty of certain mathematical problems [35]. Quantum cryptography aims to develop new cryptographic schemes that are resistant to attacks by quantum computers [36]. It makes use of the principles of quantum mechanics to provide a high level of security. Also, uses quantum mechanical properties to protect information in transit.

In traditional cryptography, the security of the system relies on the complexity of mathematical algorithms, while in quantum cryptography, the security relies on the laws of physics. Specifically, quantum cryptography uses the principle of quantum entanglement, which involves the correlation of quantum states between two particles.

The most widely known application of quantum cryptography is quantum key distribution (QKD) [37]. QKD is a protocol that enables two parties to establish a shared secret key that is completely secure against eavesdropping, even by an attacker with unlimited computing power. QKD works by transmitting a series of quantum states, or qubits, between two parties, typically named Alice and Bob. The qubits are generated using a laser and a polarizer. Alice sends a random sequence of polarizations to Bob, who measures the polarizations using his own set of polarizers. By comparing the polarizations, Alice and Bob can detect the presence of an eavesdropper.

There are many challenges to overcome before quantum cryptography can be widely adopted. One of the main challenges is the difficulty of building practical quantum cryptography systems, which require precise control of the quantum states involved. Additionally, there is a need for more research in quantum computing, as well as a need for new protocols that can be used to secure communications in different contexts.

7.2 Homomorphic encryption

Homomorphic encryption is another type of encryption that allows computation to be performed on ciphertext [38], which means that data can be encrypted and manipulated without the need to decrypt it first. In other words, it enables computations to be performed on data without revealing the data itself. This is a significant breakthrough in the field of cryptography as it allows for secure computation and data analysis without compromising privacy [39]. Homomorphic encryption has numerous applications in various fields such as finance, healthcare, and cloud computing [40]. For instance, it can be used to perform secure data analysis on sensitive data [41], such as medical records, without the need to reveal the data to unauthorized parties. It can also be used in cloud computing to protect data privacy while still allowing for secure computation in the cloud.

7.3 Block chain cryptography

Blockchain-based cryptography is a critical component of blockchain technology, which is widely used in various fields such as finance, healthcare, and supply chain management [42]. it is a distributed ledger that records transactions in a secure and transparent manner. Cryptography is used in blockchain to ensure the confidentiality, integrity, and authenticity of data stored in the blockchain network.

One of the essential cryptographic techniques used in blockchain is the digital signature. A digital signature is a mathematical scheme that validates the authenticity and integrity of a message or data. Digital signatures are used to verify transactions in the blockchain network, ensuring that the sender is the actual owner of the assets and preventing any tampering of the data [42].

Another critical cryptographic technique used in the blockchain is hash functions. Hash functions are used to create a unique digital fingerprint of data stored in the blockchain network. This unique digital fingerprint, also known as a hash value, ensures that the data is tamper-proof and cannot be altered without being detected.

Blockchain technology also employs public-key cryptography, which is a cryptographic technique that uses a pair of keys, one public and one private. Public keys are used to encrypt data, while private keys are used to decrypt data. This technique ensures the confidentiality and security of data stored in the blockchain network.

Blockchain-based cryptography plays a vital role in ensuring the security and transparency of data stored in the blockchain network. As blockchain technology continues to evolve, we can expect to see new cryptographic techniques and algorithms that will further enhance the security and efficiency of blockchain-based applications.

7.4 Multiparty computation

Multiparty computation (MPC) is a cryptographic technique that enables a group of parties to jointly compute a function on their private inputs, without revealing those inputs to each other or to any third party. This technique allows parties to collaborate and compute a result without sharing their individual data, which can be particularly useful in scenarios where data privacy is critical, such as in financial transactions or medical research [43].

Each party inputs its private data into the system, which then generates a shared output based on the combined inputs of all parties. The protocol ensures that no individual party can learn anything about the private inputs of any other party, and the final output is only known to those parties who have contributed inputs.

MPC has many practical applications, including secure auctions, electronic voting systems, and privacy-preserving data analysis. However, it can be computationally expensive, especially when the number of parties and the complexity of the function being computed increase. Despite these challenges, MPC is a powerful tool for achieving secure collaboration and computation among multiple parties [44].

7.5 Lightweight cryptography

Lightweight cryptography refers to a subset of cryptographic algorithms that are specifically designed to operate efficiently on low-resource devices such as smart cards, RFID tags, and wireless sensor nodes. These devices often have limited processing power, memory, and energy resources, making it challenging to implement traditional cryptographic algorithms on them. Lightweight cryptography aims to address these challenges by developing cryptographic algorithms that have low computational and memory requirements, while still providing a reasonable level of security.

The development of lightweight cryptography has become increasingly important with the proliferation of the Internet of Things (IoT) and other low-power, low-cost devices. These devices are becoming more prevalent in our daily lives, and many of them require secure communication and authentication. Lightweight cryptography can provide a practical and efficient solution for securing these devices, without sacrificing security. Some examples of lightweight cryptography algorithms include SIMON and SPECK block ciphers, which were designed by the National Security Agency (NSA) for use in constrained environments. Another example is the lightweight version of the advanced encryption standard (AES), known as AES-Lite. These algorithms have been adopted by various standardization bodies and are widely used in industry for securing low-resource devices.

8. Conclusions

Cryptography is a critical aspect of modern information security. It has evolved significantly over time, from basic substitution ciphers to sophisticated algorithms that provide secure communication and transactions. Today, we have various types of cryptographic schemes, including symmetric and asymmetric encryption, hash functions, digital signatures, homomorphic encryption, and multiparty computation. The development of lightweight cryptography has also enabled secure communication and transactions on low-power devices such as IoT devices. As technology continues to advance, the field of cryptography will play an increasingly vital role in ensuring secure communication and transactions in an interconnected world. The future of cryptography is exciting and promising, and we can expect to see more innovations that will enhance the security and privacy of our digital world.

Author details

Monther Tarawneh
Computer Science Department, Isra University, Amman, Jordan

*Address all correspondence to: mtarawneh@iu.edu.jo

References

[1] Bruce S. Applied cryptography: protocols, algorithms, and source code in C. 2nd ed. Hoboken, New Jersey: John Wiley & Sons; 1996

[2] Diffie W, Hellman ME. Multiuser cryptographic techniques. In: Proceedings of the June 7-10, 1976, national computer conference and exposition. ACM Digital Library; 1976. pp. 109-112

[3] Blakley GR, Borosh I. Rivest-Shamir-Adleman public key cryptosystems do not always conceal messages. Computers & Mathematics with Applications. 1979; **5**:169-178

[4] Rescorla E. Diffie-Hellman Key Agreement Method. 2070-1721, 1999

[5] Sobti R, Geetha G. Cryptographic hash functions: A review. International Journal of Computer Science Issues (IJCSI). 2012;**9**:461

[6] Rogaway P, Shrimpton T. Cryptographic hash-function basics: Definitions, implications, and separations for preimage resistance, second-preimage resistance, and collision resistance. In: FSE, 2004, Lecture Notes in Computer Science. Vol. 3017. Springer Verlag; 2004. pp. 371-388

[7] Menezes AJ, van Oorschot PC, Vanstone SA. Handbook of applied cryptography (202101 ed.). 2021;**1**:1-810

[8] Wong D. Real-world cryptography. Shelter Island, NY: Manning Publications; 2021

[9] Chaubey NK, Prajapati BB. Quantum cryptography and the future of cyber security. Hershey, PA: IGI Global; 2020. DOI: 10.4018/978-1-7998-2253-0

[10] Poongothai T, Jayarajan K, Rajeshkumar G, Patra P. Blockchain technology in healthcare applications. Journal of Critical Reviews. 2020;**7**: 8701-8707

[11] Bertaccini M. Cryptography algorithms: A guide to algorithms in blockchain, quantum cryptography, zero - knowledge protocols, and omomorphic encryption. Birmingham, UK: Packt Publishing, Limited; 2022. DOI: 10.1007/978-183882-844-4

[12] Singh S. The Code Book. Vol. 7. New York: Doubleday; 1999

[13] Davies D. A brief history of cryptography. Information Security Technical Report. 1997;**2**:14-17

[14] Mendelsohn CJ. Blaise de Vigenère and the "Chiffre Carré". In: Proceedings of the American Philosophical Society. 1940;**83**(4):103-129

[15] Schrödel T. Breaking short Vigenère ciphers. Cryptologia. 2008;**32**:334-347

[16] Wade NJ. Charles Wheatstone (1802–1875). ed: SAGE Publications ed. Vol. 31. London, England: Sage UK; 2002. pp. 265-272

[17] Kruh L. Cipher equipment. Cryptologia. 1977;**1**:143-149

[18] Smart NP, Smart NP. The enigma machine. Cryptography Made Simple. 2016;**64**(2):133-161

[19] Sidhu A. Analyzing modern cryptography techniques and reviewing their timeline. Security and Communication Networks. 2023;**10**:1-18

[20] Stamp M. Information security: principles and practice. Hoboken, NJ: John Wiley & Sons; 2011

[21] Smid ME. Development of the advanced encryption standard. Journal of Research of the National Institute of Standards and Technology. 2021;**126**:1-18

[22] Daemen J, Rijmen V. AES proposal: Rijndael. National Institute of Standards and Technology; 1999

[23] Schneier B. Description of a new variable-length key, 64 bit block cipher (Blowfish). In: Fast Software Encryption: Cambridge Security Workshop Cambridge, UK, December 9 11, 1993 Proceedings. Berlin, Heidelberg: Springer; 2005. pp. 191-204

[24] Schneier B. The twofish encryption algorithm. Dr Dobb's Journal: Software Tools for the Professional Programmer. 1998;**23**:30-34

[25] Rivest RL. The RC4 encryption algorithm, 1992. Vol. 25. RSA Data Security Inc.; 2016. pp. 1-23.

[26] Rivest RL, Shamir A, Adleman L. A method for obtaining digital signatures and public-key cryptosystems. Communications of the ACM. 1978;**21**:120-126

[27] Hellman M. New directions in cryptography. IEEE Transactions on Information Theory. 1976;**22**:644-654

[28] Lenstra HW. Factoring integers with elliptic curves. Annals of Mathematics. 1987;**126**(3):649-673

[29] Pachghare V. Cryptography and information security. Noida, Uttar Pradesh, India: PHI Learning Pvt. Ltd.; 2019

[30] Katz J, Lindell Y. Introduction to modern cryptography. Boca Raton, FL: CRC Press; 2020

[31] Li Y. Design and analysis of cryptographic protocols [Dissertation], 2015. Bochum: Ruhr-Universität Bochum; 2016

[32] Carts DA. A review of the Diffie-Hellman algorithm and its use in secure internet protocols. SANS Institute; 2001; **751**:1-7

[33] Medina R III. Systems and Methods for Digital Signature Detection. ed: Google Patents ed. 2015

[34] Tarawneh M, AlZyoud F, Sharrab Y, Kanaker H. Secure E-health framework in cloud-based environment. In: 2022 International Arab Conference on Information Technology (ACIT). IEEE; 2022. pp. 1-5

[35] Subramani S, Svn SK. Review of security methods based on classical cryptography and quantum cryptography. Cybernetics and Systems. 2023;**54**(1):1-19

[36] Mavroeidis V, Vishi K, Zych MD, Jøsang A. The impact of quantum computing on present cryptography. arXiv Preprint arXiv:1804.00200. 2018

[37] Renner R. Security of quantum key distribution. International Journal of Quantum Information. 2008;**6**:1-127

[38] Lauter KE, Dai W, Laine K. Protecting privacy through homomorphic encryption. Cham, Switzerland: Springer; 2022

[39] Doan TVT, Messai M-L, Gavin G, Darmont J. A survey on implementations of homomorphic encryption schemes. The Journal of Supercomputing. 2023;**79**: 15098-15139

[40] Chatterjee A, Aung KMM. Fully homomorphic encryption in real world applications. Singapore: Springer; 2019

[41] Viand A, Knabenhans C, Hithnawi A. Verifiable fully homomorphic encryption. arXiv Preprint arXiv:2301.07041. 2023

[42] Bolfing A. Cryptographic Primitives in Blockchain Technology: A Mathematical Introduction. New York, USA: Oxford University Press; 2020

[43] Goldreich O. Secure multi-party computation. Manuscript. Preliminary version. 1998;**78**:1-78

[44] Darby ML, Nikolaou M. MPC: Current practice and challenges. Control Engineering Practice. 2012;**20**:328-342

Chapter 3

Securing Java Source Files Using the Cipher Cryptographic Algorithm

Asma'a Al-Hakimi, Muhammad Ibrahim Ravi Bin Gobi and Misbah ul Iman

Abstract

Reverse engineering poses a substantial threat to software and hardware systems, allowing unauthorized access to proprietary information, algorithms, and trade secrets. This chapter introduces a new cryptographic algorithm designed to thwart reverse engineering efforts. The research methodology involved conducting experiments to assess the technique's efficacy, yielding promising results. It emphasizes the importance of integrating cryptographic methods as a preventive measure against reverse engineering. Through encryption, decryption, and obfuscation, developers can bolster their systems' security and mitigate reverse engineering risks. The experiments validate the algorithm's potential in safeguarding sensitive information and intellectual property. Reverse engineering involves analyzing and reconstructing a system's design, code, or logic to gain unauthorized access or replicate its functionality, potentially leading to intellectual property theft, security breaches, and financial losses. The proposed technique significantly hampers reverse engineering attempts, making it challenging for attackers to understand the system's logic or extract meaningful information.

Keywords: cryptography, cipher algorithm, reverse engineering, software security, symmetric encryption

1. Introduction

Reverse engineering is the process of analyzing a product, system, or technology to understand its design, functionality, and operation. It involves disassembling, examining, and studying the components, algorithms, and code of the subject to gain insights into how it works. In recent years, reverse engineering has become increasingly important and has faced emerging challenges, new perspectives, and innovative applications [1].

Let us explore some of these aspects.

- Intellectual Property Protection: Reverse engineering can be used to gain unauthorized access to proprietary information, leading to concerns about intellectual property theft. Protecting trade secrets and maintaining the legal boundaries of reverse engineering is a significant challenge [2].

- Digital Rights Management (DRM): Reverse engineering is often used to circumvent DRM mechanisms employed by software, games, or digital media. Developers face the challenge of creating robust DRM solutions that can resist reverse engineering attempts.

- Obfuscation and Anti-Reverse Engineering Techniques: To protect software and prevent reverse engineering, developers employ various obfuscation techniques. These techniques make the code more complex and harder to understand, posing challenges to reverse engineers [3, 4].

- Security Analysis and Vulnerability Discovery: Reverse engineering is crucial in identifying security vulnerabilities in software, firmware, or hardware. By analyzing these systems, experts can uncover potential weaknesses and provide insights to improve security.

- Legacy Systems and Interoperability: Reverse engineering plays a vital role in understanding legacy systems that lack proper documentation or have become obsolete. Reverse engineering enables integration with modern technologies and facilitates interoperability [5].

- Malware Analysis: Reverse engineering is crucial in analyzing and understanding malware, such as viruses, worms, and Trojans. Security professionals can reverse engineer malicious code to uncover their behavior, techniques, and potential countermeasures [6].

- Product and Process Improvement: Reverse engineering helps in gaining insights into competitors' products, which can be used to enhance one's own products or develop innovative solutions. By reverse engineering products, companies can identify shortcomings, understand manufacturing processes, and find areas for improvement.

- Patent Infringement Investigations: Reverse engineering can be used to investigate potential patent infringement cases. By examining the design and functionality of a product, experts can determine if it violates existing patents, aiding legal proceedings [7].

- 3D Printing and Manufacturing: Reverse engineering is widely used in 3D scanning and modeling to replicate or modify physical objects. By scanning and analyzing existing objects, manufacturers can create accurate digital models for reproduction or redesign [8].

- Software Maintenance and Documentation: Reverse engineering can be employed to understand and document legacy software systems. It helps software developers comprehend codebases that lack proper documentation, aiding in maintenance, bug fixing, and system upgrades [9].

2. Exploration of existing cryptography techniques

Cryptography is a fundamental aspect of modern computer security and plays a crucial role in protecting sensitive information, ensuring data integrity, and enabling secure communication over insecure channels [10].

Cryptography is the practice and study of techniques used to secure information and communication by converting it into an unintelligible form, thereby protecting it from unauthorized access or malicious activities. It is an essential aspect of modern information security, providing confidentiality, integrity, authentication, and non-repudiation [11].

The primary objective of cryptography is to ensure that sensitive data remains confidential during storage, transmission, and processing. This is achieved using cryptographic algorithms, which manipulate the plaintext (**original message**) into ciphertext (**encrypted message**) using encryption keys. The encrypted data can only be deciphered back to its original form by authorized individuals who possess the corresponding decryption keys [12].

There are two fundamental categories of cryptography: symmetric key cryptography and public key cryptography. In symmetric key cryptography, a single key is used for both encryption and decryption. This key must be securely shared between the sender and the intended recipient. On the other hand, public key cryptography employs a pair of mathematically related keys: a public key for encryption and a private key for decryption. The public key can be freely distributed, while the private key must be kept secret [13].

Cryptography techniques are used in various applications, such as secure communication protocols (e.g., SSL/TLS), data encryption at rest and in transit, digital signatures, secure authentication systems, and secure storage of sensitive information. It plays a crucial role in protecting personal data, financial transactions, confidential business information, and government secrets.

The strength of cryptographic systems relies on the complexity of the algorithms and the length of the encryption keys used. Cryptanalysis is the science of analyzing cryptographic systems to identify vulnerabilities or weaknesses that could be exploited by attackers. Consequently, ongoing research and development in cryptography are essential to stay ahead of potential threats and ensure the robustness of security mechanisms [14].

In recent years, emerging technologies such as blockchain, quantum computing, and homomorphic encryption have introduced new challenges and opportunities in the field of cryptography. These advancements continue to drive the evolution of cryptographic techniques to address the changing landscape of information security and privacy [15].

The following section presents cryptography techniques:

2.1 Symmetric encryption

Symmetric encryption algorithms use a single shared key to both encrypt and decrypt data. Well-known symmetric encryption algorithms include the Advanced Encryption Standard (AES), which is widely used and considered secure. AES operates on fixed-size blocks (128 bits) and supports key sizes of 128, 192, or 256 bits [16].

2.2 Asymmetric encryption

Asymmetric encryption (also known as public key encryption) employs a pair of mathematically related keys: a public key for encryption and a private key for decryption. The encryption key is made public, allowing anyone to encrypt messages, while the decryption key remains private. Popular asymmetric encryption algorithms include Rivest-Shamir-Adleman (RSA) (encryption technique) and Elliptic Curve Cryptography (ECC) [17].

2.3 Hash functions

Hash functions take an input (message) and produce a fixed-size output (hash value). A key feature of hash functions is that they are one-way, meaning it is computationally infeasible to derive the original message from the hash value. Commonly used hash functions include the Secure Hash Algorithm (SHA) family, such as SHA-256, SHA-384, and SHA-512 [18].

2.4 Digital signatures

Digital signatures are used to verify the authenticity and integrity of digital documents or messages. They involve the use of asymmetric encryption to create a unique signature for a document that can be verified by anyone with access to the signer's public key. The Digital Signature Algorithm (DSA) and the Elliptic Curve Digital Signature Algorithm (ECDSA) are widely used for digital signatures [19].

2.5 Key exchange protocols

Key exchange protocols establish a shared secret key between two parties communicating over an insecure channel. One widely used key exchange protocol is the Diffie-Hellman key exchange, which allows two parties to generate a shared secret without explicitly transmitting it over the network [20].

2.6 Post-quantum cryptography

With the rise of quantum computing, there has been increasing interest in post-quantum cryptography. These cryptographic algorithms are designed to be resistant to attacks by quantum computers. Examples include lattice-based cryptography, code-based cryptography, and multivariate cryptography [21].

Cryptography is a dynamic field, and new techniques and algorithms are constantly being developed to address emerging threats and technological advancements. Researchers and practitioners are continually working to improve cryptographic techniques to ensure the security of digital communications and protect sensitive information [22].

3. The mechanism of cryptography

The mechanism of cryptography to protect code can be represented conceptually through a series of steps. Here is a graphical representation using text:

A. Original Code: Represents the source code that needs to be protected.

B. Encryption: The original code is input into an encryption algorithm along with a secret encryption key. The encryption algorithm transforms the original code into ciphertext.

C. Ciphertext: The encrypted form of the original code. It appears as random and unreadable data.

D. Decryption Key: The secret decryption key is needed to reverse the encryption process and retrieve the original code.

E. Decryption: The ciphertext, along with the decryption key, is input into a decryption algorithm. The decryption algorithm reverses the encryption process and retrieves the original code.

F. Authorized Access: The original code is now accessible and readable by authorized individuals who possess the decryption key.

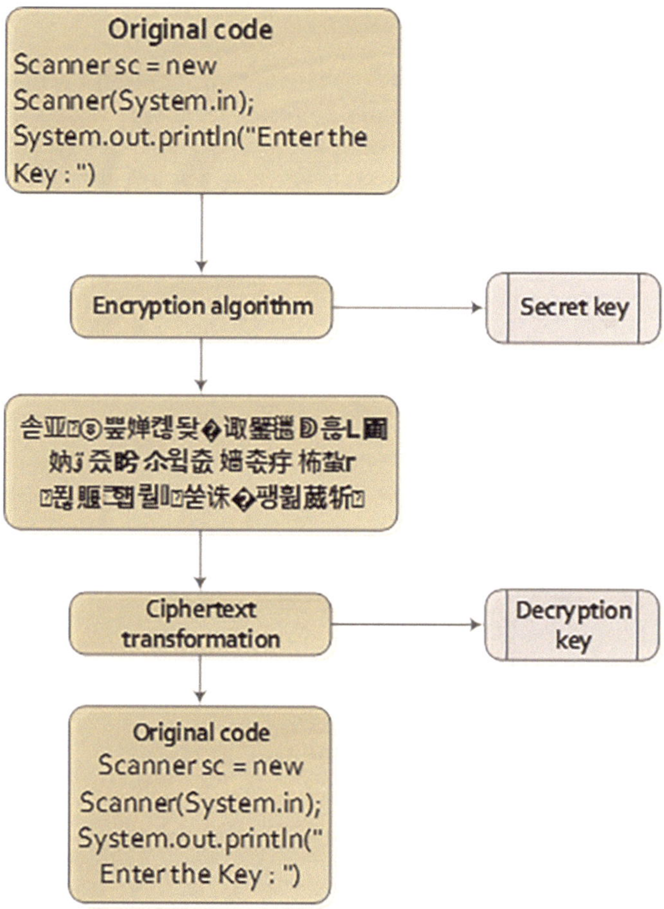

Figure 1.
Cryptographic algorithm process.

While this textual representation does not provide a visual graph, it outlines the sequential steps involved in the mechanism of cryptography to protect code. The encryption step converts the original code into ciphertext, making it unreadable without the decryption key. The decryption step, performed by authorized individuals with the decryption key, retrieves the original code for access and use. **Figure 1** illustrates the encryption process.

4. Distinguishing symmetric encryption from asymmetric encryption

4.1 Symmetric encryption

Symmetric encryption, also known as secret-key encryption, uses a single shared key to both encrypt and decrypt data. The same key is used by both the sender and the receiver, which means that both parties must have access to the key in advance. The strength and security of symmetric encryption lie in the secrecy of the shared key [23].

4.1.1 Case 1

One of the most widely used symmetric encryption algorithms is the Advanced Encryption Standard (AES). AES operates on fixed-size blocks of data and supports key sizes of 128, 192, or 256 bits. For example, if Eva wants to send an encrypted message to YG using AES, they both need to agree on a secret key beforehand. EVA encrypts the message using the shared key and sends the encrypted ciphertext to YG. YG then uses the same shared key to decrypt the ciphertext and retrieve the original message.

4.1.2 Case 2

```
import javax.crypto.Cipher;
import javax.crypto.KeyGenerator;
import javax.crypto.SecretKey;
import javax.crypto.spec.SecretKeySpec;
import java.nio.charset.StandardCharsets;
import java.util.Base64;
public class SymmetricEncryptionExample
{
  public static void main(String[] args) throws Exception
   {
 // Generate a random AES key
 SecretKey secretKey = generateRandomAESKey();
 // Create the AES cipher instance
     Cipher cipher = Cipher.getInstance("AES");
 // Set the cipher to encrypt mode using the generated key
 cipher.init(Cipher.ENCRYPT_MODE, secretKey);
 // The plaintext message to be encrypted
 String plaintext = "This is a secret message!";
```

```
// Convert the plaintext to bytes
byte[] plaintextBytes = plaintext.getBytes(StandardCharsets.UTF_8);
// Encrypt the plaintext
byte[] ciphertextBytes = cipher.doFinal(plaintextBytes);
// Convert the ciphertext to Base64 for easier display
String ciphertext = Base64.getEncoder().encodeToString(ciphertext
 Bytes);
// Print the encrypted ciphertext
System.out.println("Ciphertext: " + ciphertext);
 }
private static SecretKey generateRandomAESKey() throws Exception
{
// Generate a 256-bit AES key
KeyGenerator keyGen = KeyGenerator.getInstance("AES");
keyGen.init(256);
return keyGen.generateKey();
 }
}
```

Output

```
Ciphertext: ZH6cLGRamKYTn/44XovM7kMMqVITVEb6GUsVr1ZXZGk=
```

4.2 Asymmetric encryption

Asymmetric encryption, also known as public key encryption, employs a pair of mathematically related keys: a public key and a private key. The public key is made freely available, while the private key must be kept secret. These keys are mathematically linked, but it is computationally infeasible to derive the private key from the public key. **Table 1** presents the comparison of Symmetric and Asymmetric in terms of key security.

4.2.1 Case 1

The RSA algorithm is a widely used asymmetric encryption algorithm. In RSA, each user has a pair of keys: a public key and a private key. For example, if EVA wants

	Symmetric	Asymmetric
Key Management	Symmetric encryption requires secure key distribution [24].	Asymmetric encryption eliminates this need. Asymmetric encryption simplifies key management by using public and private key pairs.
Computational Complexity	Symmetric encryption algorithms are generally faster and more efficient computationally, making them suitable for encrypting large amounts of data.	Asymmetric encryption algorithms are slower and computationally more expensive.
Security	Both symmetric and asymmetric encryption provide security, but the level of security differs. Symmetric encryption relies on the secrecy of the shared key [25].	While asymmetric encryption relies on the computational difficulty of deriving the private key from the public key.

Table 1.
Comparison between symmetric and asymmetric.

to send an encrypted message to YG, she can use YG's public key to encrypt the message. Only YG, with the corresponding private key, can decrypt the message.

4.2.2 Case 2

```
import java.security.KeyPair;
import java.security.KeyPairGenerator;
import java.security.PrivateKey;
import java.security.PublicKey;
import java.security.SecureRandom;
import javax.crypto.Cipher;
import java.util.Base64;
public class AsymmetricEncryption
{
    public static void main (String[] args) throws Exception
    {
        // Generate key pair (public and private keys)
      KeyPair keyPair = generateKeyPair();
      // Get the public and private keys from the key pair
      PublicKey publicKey = keyPair.getPublic();
      PrivateKey privateKey = keyPair.getPrivate();
      // Create the RSA cipher instance
      Cipher cipher = Cipher.getInstance("RSA");
      // Set the cipher to encrypt mode using the public key
      cipher.init(Cipher.ENCRYPT_MODE, publicKey);
      // The plaintext message to be encrypted
      String plaintext = "This is a secret message!";
      // Convert the plaintext to bytes
      byte[] plaintextBytes = plaintext.getBytes();
      // Encrypt the plaintext
      byte[] ciphertextBytes = cipher.doFinal(plaintextBytes);
      // Convert the ciphertext to Base64 for easier display
      String ciphertext = Base64.getEncoder().encodeToString(cipher
       textBytes);
      // Print the encrypted ciphertext
      System.out.println("Ciphertext: " + ciphertext);
    }
  private static KeyPair generateKeyPair() throws Exception
  {
        // Generate an RSA key pair with a key size of 2048 bits
        KeyPairGenerator keyGen = KeyPairGenerator.getInstance("RSA");
        keyGen.initialize(2048, new SecureRandom());
        return keyGen.generateKeyPair();
  }
}
```

Output

Ciphertext:
oBkZMzjPbFAl5IRBtT6nOg4B6AeqQ0sYZhI06zHG6s1Qm1n49kZZlU8ZJ/sTf8BSyY7ZCBtkzVK/92vX6aRnLP4wW6qYr6G1HzvrVMu2U6zsgl0s6FBXCtd44/WJ/Ds/xCnS9Q1Pq5u1Ly2ePXjNDEZ79WkLbLJqfbr9P4YSBE=

5. Cryptographic algorithm to prevent reverse engineering for Java source file

5.1 Original code

This section discusses the method of implementing the cryptographic algorithm to prevent reverse engineering of Java applications. The algorithm shall be deployed in the source file of the Java application. In this section, the first discussion is a presentation of implementing the cryptographic algorithm, and the second discussion is a presentation of methodology testing of the algorithm's effectiveness in stopping reverse engineering.

5.1.1 Implementing cryptographic algorithm

To prevent reverse engineering, usually, protection is applied to the class file while the source file is left without protection. In this chapter, the protection will be in the source file. The cryptographic algorithms that will be applied for the protection are AES and Cipher. The following steps are to discuss the flow of protecting the code.

Generate the AES secret key. It is optional to create it mathematically or using KeyGenerator; for the purpose of this research, it will be created mathematically to ensure more security.

Apply the Cipher algorithm. To apply the Cipher algorithm for encryption, Java Cryptography Architecture (JCA) and Java Cryptography Extension (JCE) libraries. These libraries provide very important information to escalate the security of Java applications.

Create SecretKey as an object; the secretkey class presents a secretkey for an encryption algorithm.

Initialize the cipher object; the cipher class will be used for encryption and decryption.

After applying the cipher and secretkey for the encryption, perform encryption operation, after initializing the cipher object and the secretkey.

The cipher and the Java Cryptography Architecture will absorb the code in the source file to be encrypted, then while compiling the source to create the binary file, an extra layer of encryption is created, which will add more security to the application. In the case of decryption, the code will be transformed to different language that is not readable by human. However, the application is running perfectly and providing the required output.

5.1.2 Testing methodology

In this section, empirical evaluation consists of an experiment. The measurements of the experiment are the ability to execute the code to provide the same output as the original code and the ability to read variables, classes, and logic flow of the code. The environment used to conduct the experiment is NetBeans. The attributes of the experiment are lines of code (LOC) before and after reversing, the number of libraries created after reversing and comparing against the original code, the number of discovered methods, and the comparison against the original code.

The tools used for the reversing against the cryptographic algorithm are JD and CAVAJ reversing tools to determine the ability of the cryptographic to protect the source file from reversing. **Figure 2** illustrates the structure of the experiment.

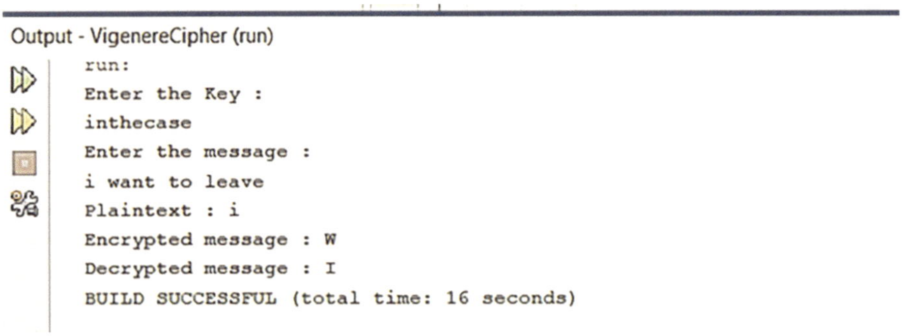

Figure 2.
Experiment design.

To start the process of the experiment, the original code of Java must be provided, and the output of the running program must be able to be compared after applying the cryptography algorithm. The following steps present the process of deploying the algorithm to the original code.

A. Step 1: Get the Java original code in the source file. The following code is Java source file.

```
import java.util.Scanner;
public class VigenereCipher {
 public static String encrypt (String text, final String key)        {
      String res = "";
      text = text.toUpperCase();
 for (int x = 0, y = 0; x < text.length(); x++)           {
         char c = text.charAt(x);
 if (c < 'A' || c > 'Z')
 continue;
         res += (char) ((c + key.charAt(y) - 2 * 'A') % 26 + 'A');
 y = ++ y % key.length();                                 }
      return res;                                                    }
 public static String decrypt(String text, final String key)        {
       String res = "";
      text = text.toUpperCase();
      for (int x = 0, y = 0; x < text.length(); x++)      {
         char c = text.charAt(i);
         if (c < 'A' || c > 'Z')
 continue;
         res += (char) ((c - key.charAt(y) + 26) % 26 + 'A');
 y = ++ y % key.length();                                 }
      return res;                                                    }
 public static void main(String[] args)                              {
      Scanner sc = new Scanner(System.in);
```

```
System.out.println("Enter the Key : ");
      String key = sc.next();
      Scanner sc1 = new Scanner(System.in);
System.out.println("Enter the message : ");
      String message = sc.next();
      String encryptedMsg = encrypt(message, key);
System.out.println("Plaintext : " + message);
System.out.println("Encrypted message : " + encryptedMsg);
System.out.println("Decrypted message : " + decrypt(encryptedMsg, key));    }}
```

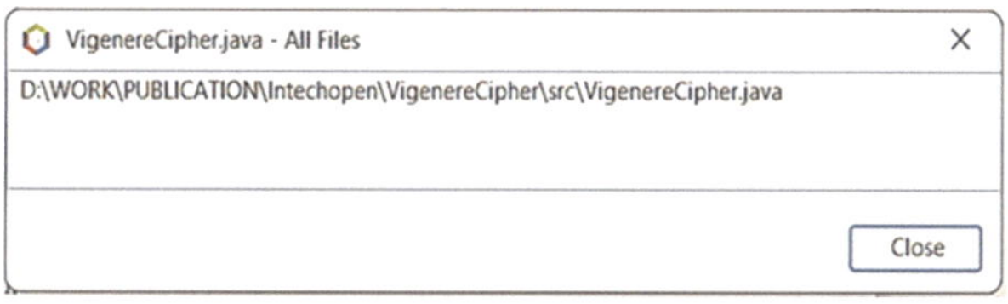

Figure 3.
Output of original code.

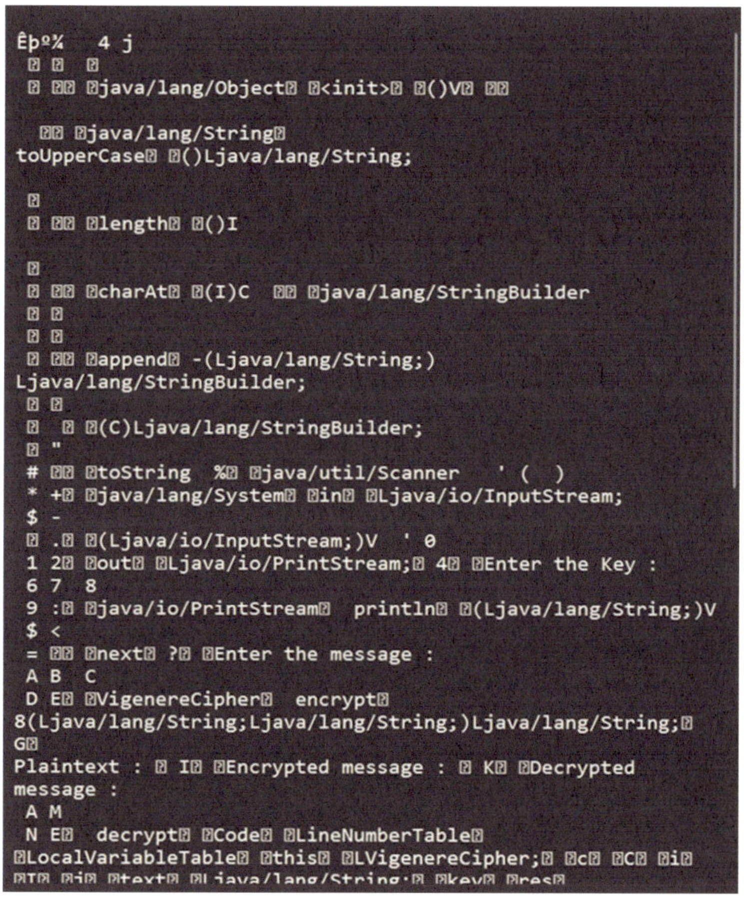

Figure 4.
Original file location.

The above code is to get input from the user and provide feedback based on the input required. The reason for using a communicative program is to ensure that after applying the algorithm, the program still has the ability to perform and provide proper and understandable communication to the user and to make sure that the algorithm did not change or harm the structure of the original code. **Figure 3** illustrates the output of the original code. **Figure 4** illustrates the location of the file created after running the application.

Based on the above running code and the output, this will be the benchmark to be used for experimenting on the code after deploying the algorithm to make sure that the program is sufficient and usable and provide proper output that is understandable by the user.

B. Step 2: Deploying cryptographic algorithms to protect source code. **Figure 4** illustrates the class file of cryptographic algorithm.

```
import java.io.IOException;
import javax.crypto.Cipher;
import javax.crypto.spec.SecretKeySpec;
import java.nio.charset.StandardCharsets;
import java.nio.file.Files;
import java.nio.file.Path;
import java.nio.file.Paths;
import java.security.InvalidKeyException;
import java.security.NoSuchAlgorithmException;
import java.util.Base64;
import javax.crypto.BadPaddingException;
import javax.crypto.IllegalBlockSizeException;
import javax.crypto.NoSuchPaddingException;
public class JavaSourceCodeEncryption     {
    public static void main(String[] args) throws IOException, InvalidKeyException,
    IllegalBlockSizeException, BadPaddingException, NoSuchPaddingException,
    NoSuchAlgorithmException     {
 String sourceCodePath =("D:\\WORK\\PUBLICATION\\Intechopen\\
 VigenereCipher\\src\\VigenereCipher.java");
 String encryptionKey = "ClosingKey";
 // Read the Java source code file
 String sourceCode = readSourceCodeFromFile(sourceCodePath);
 // Encrypt the source code
 String encryptedCode = encryptSourceCode(sourceCode, encryptionKey);
     // Save the encrypted code to a file
 String encryptedCodePath =("D:\\WORK\\PUBLICATION\\Intechopen\\
 VigenereCipher\\src\\encrypt.txt");
 saveEncryptedCodeToFile(encryptedCode, encryptedCodePath);
 System.out.println("Java source code has been encrypted and saved as " +
 encryptedCodePath);                    }
  private static String readSourceCodeFromFile(String filePath) throws IOException
  //throws Exception                     {
  Path path = Paths.get(filePath);
  byte[] sourceCodeBytes = Files.readAllBytes(path);
  return new String(sourceCodeBytes, StandardCharsets.UTF_8);                    }
```

```
private static String encryptSourceCode(String sourceCode, String encryptionKey)
 throws InvalidKeyException, IllegalBlockSizeException, BadPaddingException,
NoSuchPaddingException, NoSuchAlgorithmException //throws Exception    {
Cipher cipher;
    cipher = Cipher.getInstance("AES/ECB/PKCS5Padding");
SecretKeySpec keySpec = new
 SecretKeySpec(encryptionKey.getBytes(StandardCharsets.UTF_8), "AES");
cipher.init(Cipher.ENCRYPT_MODE, keySpec);
byte[] encryptedBytes = cipher.doFinal(sourceCode.getBytes(StandardCharsets.
UTF_8));
return Base64.getEncoder().encodeToString(encryptedBytes); }
private static void saveEncryptedCodeToFile(String encryptedCode, String filePath)
 throws IOException    {
byte[] encryptedBytes = Base64.getDecoder().decode(encryptedCode);
   Path path = Paths.get(filePath);
Files.write(path, encryptedBytes); } }
```

Based on the above code, **Table 2** presents the description of each element.

After deploying the cryptographic algorithm into the code, the source code has been compiled to create the binary file. With the extra security layer that has been added to the class file after compiling, the code form in the class file has changed and become more encrypted, unlike the usual code. **Figure 5** illustrates the byte code presentation after applying the cryptographic algorithm.

C. Step 3: Execute and evaluate the cryptographic algorithm within the provided codebase to generate a safeguarded source file. **Figure 6** depicts the successful execution of the code for text file generation. The encrypted code is showcased in **Figure 6** after the application of the cryptography algorithm.

Element	Description
Cipher	This object is initialized with AES, ECB, PKCS5Padding transformation. AES is used for encryption and decryption as well. This method accepts the source code and encryption key as an argument and then encrypts the source code bytes encoded in Base64. From this object will be calling 'init' method associated with 'Cupher. ENCRYPT_MODE', the 'SecretKeySpec', and the plaintext bytes to be encrypted.
Secretkeyspec	This object is created to present the secret key used for the encryption and decryption process. The secret key is specified as byte array with AES algorithm.
doFinal	This method will be called on the 'Cipher' to perform the encryption. The plaintext bytes are passed as an argument to this method. The encrypted bytes are then returned as the results of the encryption operation.
saveEncryptedCodeToFile	This method is used to save the encrypted code to a file. It takes the encrypted code that is first decoded from Base64 to a byte array. Then, the byte array is written to the specified file path using the 'File.write'.
File.write	This method is meant to receive and save the byte array generated from the decoded code.

Table 2.
Encryption code description.

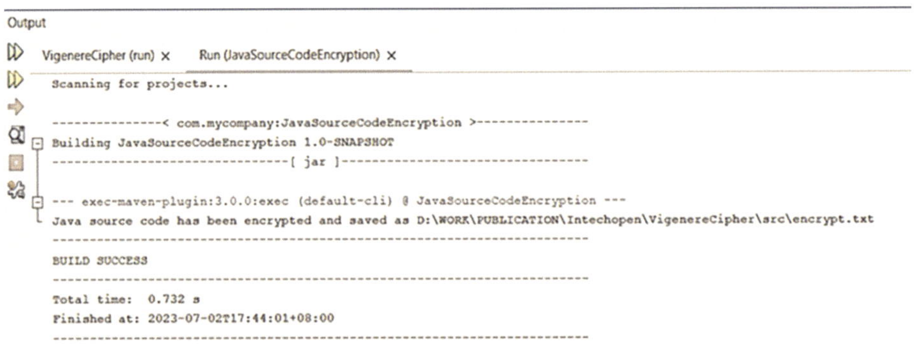

Figure 5.
Class file of cryptographic algorithm

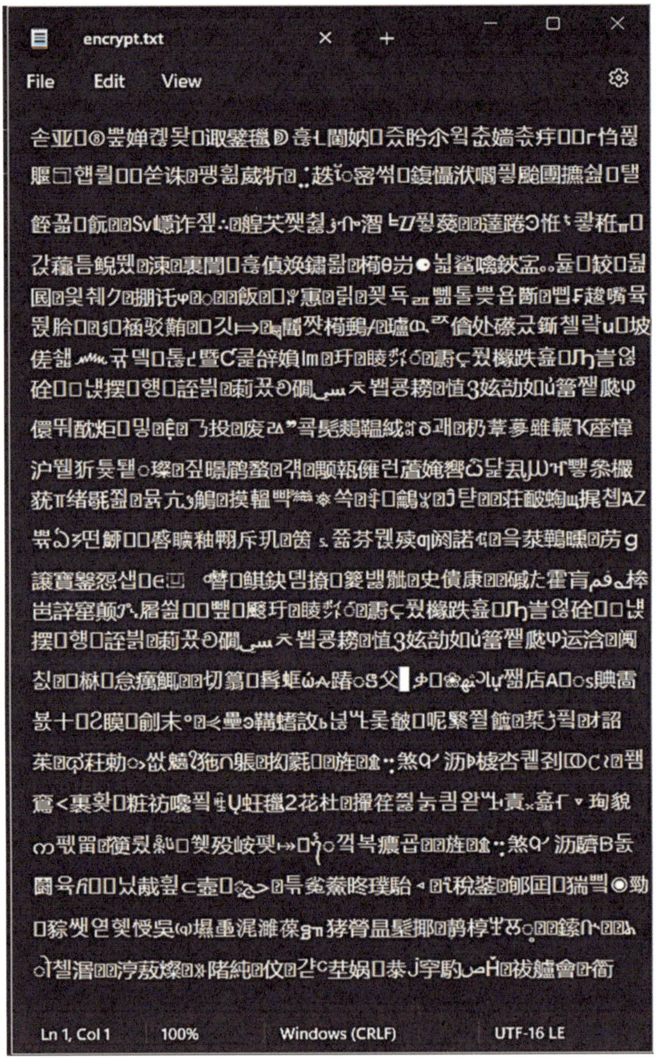

Figure 6.
Successfully created text file.

The source file has been compiled successfully and was written in the file. While reversing and revealing the source file, the original code has changed to a different one that is not readable by human; however, the application is running and providing sufficient output as required by the user; the source file is presented in **Figure 7.**

D. Step 4: Two reversing tools have been used for the experiment to test the effectiveness and efficiency of the proposed algorithm. The purpose of using JD and CAVAJ is to test with each tool the ability to reveal and read the encrypted code, test if the decrypted code is running, and provide an output that is the same as the original and encrypted code. The first tool used was CAVAJ. This tool generated an error notification as it was not able to read the encrypted class file, as presented in **Figure 8.**

The encrypted code was tested to generate output similar to the output generated from the original code; after testing, it was obvious the algorithm did not harm the code and kept it protected and provided the same output as the original code with the same amount of the time as the original code. **Figure 9** illustrates the output of the encrypted code.

Based on the testing with CAVAJ, the results were successful; the algorithm was able to defend and hide the codes from the reversing tool, the JD reversing tool was used to conduct the same test against the encrypted code, the class file was added to the environment of JD to determine the ability to reveal the code the result of the test is illustrated in **Figure 10.**

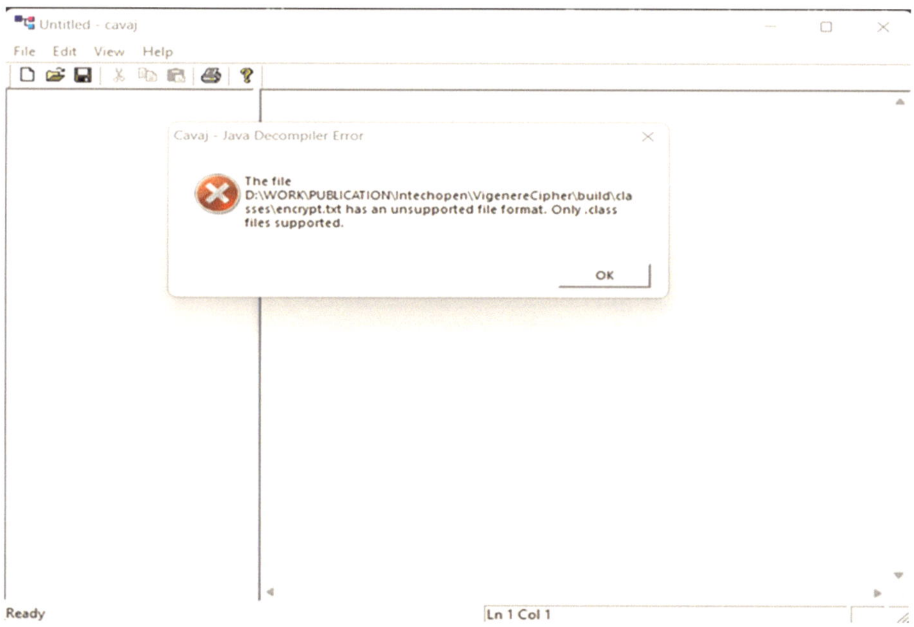

Figure 7.
Code after running the cryptography algorithm.

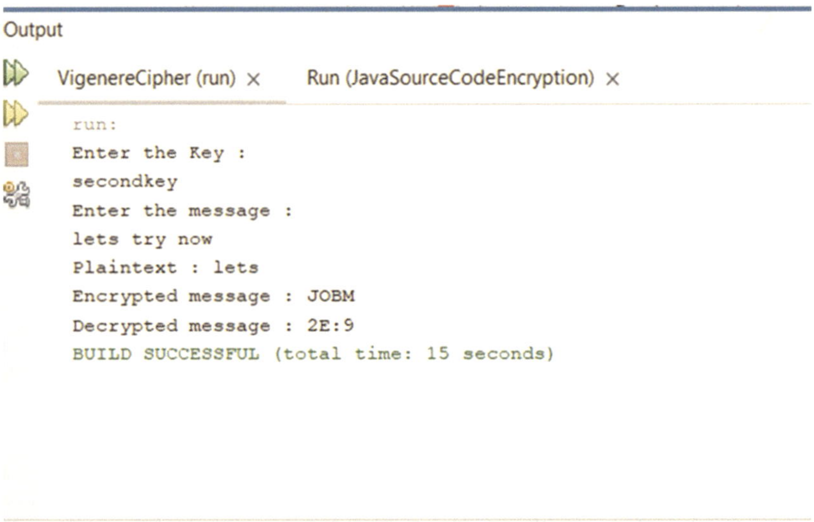

Figure 8.
Reversing encrypted source file with CAVAJ.

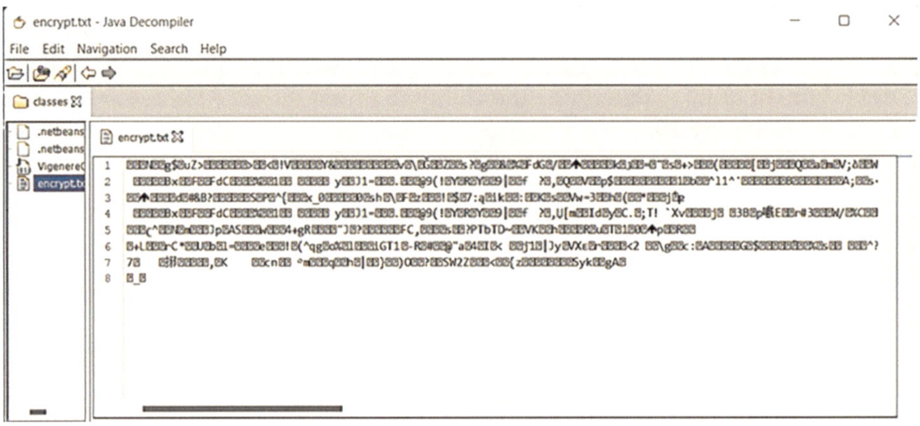

Figure 9.
Running source file after encryption.

6. Discussion and findings

Based on the experiment, the effectiveness of employing a cipher cryptographic algorithm to secure Java source files was evaluated. The experiment involved utilizing the CAVAJ and JD decompilers to ascertain whether encrypted code could be deciphered, thereby compromising its security.

The results of the experiment indicate that the encrypted code generated through the cipher cryptographic algorithm remained impervious to decryption attempts by the decomplication tools. Despite using these tools to inspect the encrypted code, no readable output was obtained; instead, error messages were encountered, signifying the robustness of the encryption. This outcome underscores the efficacy of utilizing cipher cryptographic algorithms to safeguard Java source files from unauthorized access and reverse engineering.

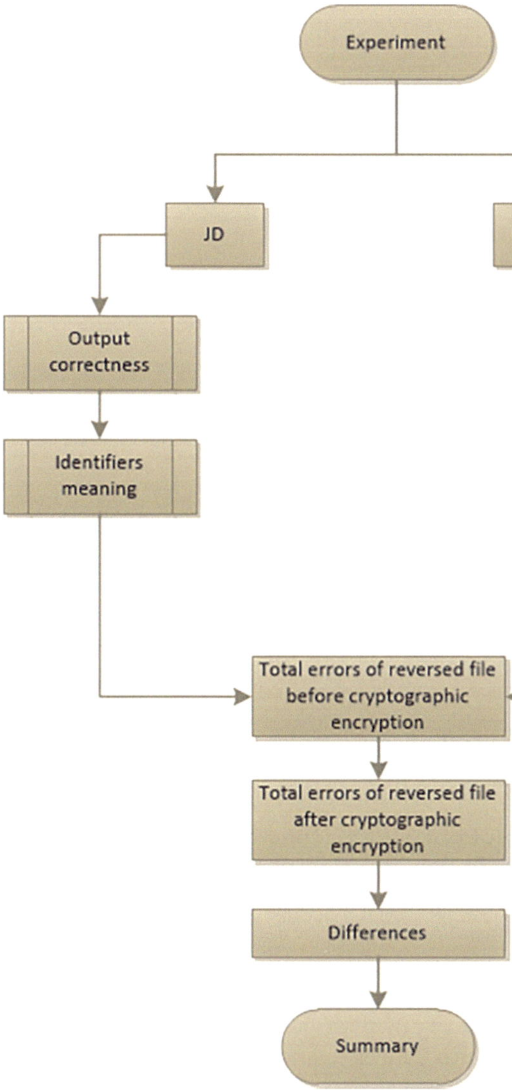

Figure 10.
Reversing encrypted source file with JD.

CAJAV reversing tool		
	Original code	**Reverse encrypted code**
LOC	422	478
Number of libraries	14	6
Number of methods	5	3
CAJAV reversing tool		
LOC	422	148

CAJAV reversing tool		
	Original code	**Reverse encrypted code**
Number of libraries	14	4
Number of methods	5	3

Table 3.
Similarity calculation before and after reversing.

Reversing tool	CAJAV		JAD	
	Parameters			
	No	Yes	No	Yes
Output correctness	✓		✓	
Syntax		✓	✓	
Reversed code error		✓		✓
Flow		✓		✓
Identifiers	✓		✓	
Methods and classes			✓	
Decrypt string test	✓			✓

Table 4.
Reversing tools correctness tests.

Furthermore, the functionality of the encrypted code was assessed to ensure that its execution remained intact post-encryption. The experiment confirmed that the encrypted code was executed with the same level of functionality as the original unencrypted code, thereby validating the suitability of the chosen cryptographic approach for preserving both security and functionality. **Table 3** presents the numerical findings of the experiment comparing encrypted and original code.

These findings have significant implications for software developers and organizations seeking to enhance the security of their Java applications. By integrating cipher cryptographic algorithms into their development processes, developers can mitigate the risk of intellectual property theft and unauthorized access to sensitive code. Additionally, the ability of encrypted code to maintain its functionality ensures that security measures do not compromise the performance or usability of the software. The ability of the reversing tool to read the encrypted code is presented in **Table 4**.

The experiment demonstrates the effectiveness of employing cipher cryptographic algorithms as a viable strategy for securing Java source files. Moving forward, further research could explore the performance of different cryptographic algorithms and their impact on both security and code execution to provide comprehensive insights for software developers and security practitioners.

7. Conclusion

The Cryptographic Algorithm to Prevent Reverse Engineering for Java Source Files has shown promising results in protecting sensitive code from unauthorized access.

Through the conducted experiment, it was observed that the proposed cryptography technique successfully encrypted the Java source file, thereby mitigating the risk of reverse engineering. The need to protect Java source files arises from the desire to safeguard intellectual property, proprietary algorithms, and confidential information embedded within the code. Reverse engineering poses a significant threat as it allows unauthorized individuals to analyze, modify, or replicate the source code, potentially compromising its integrity, security, and commercial value.

By employing cryptographic algorithms, the code can be transformed into an encrypted format that is computationally infeasible to decipher without the corresponding decryption key. While the experiment demonstrated the efficacy of the employed cryptography algorithm, it is important to acknowledge that encryption alone may not provide absolute protection. Additional security measures, such as access controls, secure development practices, and code obfuscation, should be considered in conjunction with encryption to create a more robust defense against reverse engineering. Furthermore, the proposed cryptography algorithm can be enhanced in the future by incorporating advanced encryption techniques, employing stronger key management practices, and adapting to emerging cryptographic standards.

Ongoing research and development in the field of cryptography can help address potential vulnerabilities, improve encryption algorithms, and enhance the overall security of Java source files. In summary, the Cryptographic Algorithm to Prevent Reverse Engineering demonstrates a promising approach to protect Java source files. By combining encryption with other security measures, developers can enhance the security posture of their code and mitigate the risk of unauthorized access and reverse engineering. Continued advancements in cryptography will contribute to the evolution of stronger and more effective protection mechanisms for Java source files and other valuable intellectual property.

8. Future work

Based on the positive outcomes of the cryptography algorithm proposed, there is an opportunity to develop a hybrid encryption approach to secure both the source code and class files. The algorithm has the potential to be further improved and transformed into a tool that allows developers to selectively encrypt specific sections of the code within designated methods, thereby augmenting the overall protection. Cipher algorithms can be enhanced further to explore and implement more advanced cryptographic algorithms to further enhance the security of Java source files. Evaluate the performance and effectiveness of algorithms such as AES, RSA, or others in the context of Java source file protection. Furthermore, to optimize the performance of the cryptographic algorithm to minimize its impact on the overall system performance. Consider techniques such as parallel processing, algorithmic improvements, or hardware acceleration to achieve efficient encryption and decryption.

Author details

Asma'a Al-Hakimi[1*], Muhammad Ibrahim Ravi Bin Gobi[2] and Misbah ul Iman[3]

1 Faculty of Information Sciences and Engineering, Management, and Science, University, University Drive, Off Persiaran Olahraga, Shah Alam, Selangor Darul, Ehsan, Malaysia

2 School of Computing, Faculty of Computing and Engineering, Quest International University, Ipoh, Perak, Malaysia

3 Sr. Specialist Applications, Dubai Holding Asset Management, Dubai, UAE

*Address all correspondence to: asmaa@msu.edu.my

References

[1] Luoma-aho M. Java Script Web Cryptography API. Metropolia; 2015

[2] Kessler GC. Introduction on to cryptography Webinar InformaBon. 2012 [Online]. Available from: https://commons.erau.edu/db-security-studiespartofthecommunicationtechnologyandnewmediacommons%0Ascholarly

[3] Al-Hakimi A, Md Sultan AB. Hybrid obfuscation of encryption. Coding Theory Essentials [Working Title]. IntechOpen; 2023. DOI: 10.5772/intechopen.109662

[4] Al-Sanjary OI, Ibrahim OA, Sathasivem K. A new approach to optimum steganographic algorithm for secure image. In: 2020 IEEE Int. Conf. Autom. Control Intell. Syst. I2CACIS 2020 – Proc. IEEEExplore; 2020. pp. 97-102. DOI: 10.1109/I2CACIS49202.2020.9140186

[5] Prema G, Natarajan S. Steganography using genetic algorithm along with visual cryptography for wireless network application. In: 2013 Int. Conf. Inf. Commun. Embed. Syst. ICICES. IEEEExplore; 2013. pp. 727-730. DOI: 10.1109/ICICES.2013.6508373

[6] Rao TVN. Application of elliptical curve cryptography in empowering cloud data security. International Journal of Recent and Innovation Trends in Computer Communication. 2017;**5** (August):117-121. [Online]. Available from: http://www.ijritcc.org

[7] Bulat R, Ogiela MR. Personalized cryptography algorithms – A comparison between classic and cognitive methods. In: Proc. – 52nd Annu. IEEE/IFIP Int. Conf. Dependable Syst. Networks - Suppl. Vol. DSN-S 2022. IEEEExplore; 2022. pp. 43-44. DOI: 10.1109/DSN-S54099.2022.00026

[8] Sravya G, Kumar MOVP, Sudarsana Reddy Y, Jamal K, Mannem K. The ideal block ciphers-correlation of AES and PRESENT in cryptography. In: Proc. 3rd Int. Conf. Intell. Sustain. Syst. ICISS 2020. IEEEExplore; 2020. pp. 1107-1113. DOI: 10.1109/ICISS49785.2020.9315883

[9] Matching T, Test R. Cryptography. TEST. 2001;**1937**(1992):162-173

[10] Jorstad N, Landgrave T. Cryptographic algorithm metrics. 20th National Information Systems Security. Institute for Defense Analyses Science and Technology Division; 1997. [Online]. Available from: http://csrc.nist.gov/nissc/1997/proceedings/128.pdf

[11] Kubba ZMJ, Hoomod HK. A hybrid modified lightweight algorithm combined of two cryptography algorithms PRESENT and Salsa20 using chaotic system. In: 1st Int. Sci. Conf. Comput. Appl. Sci. CAS 2019. 2019. pp. 199-203. DOI: 10.1109/CAS47993.2019.9075488

[12] Gowda SN. Innovative enhancement of the Caesar cipher algorithm for cryptography. In: Proc. – 2016 Int. Conf. Adv. Comput. Commun. Autom. (Fall), ICACCA 2016. 2016. pp. 1-4. DOI: 10.1109/ICACCAF.2016.7749010

[13] Hunacek M. Cryptography: An introduction. In: Introductin to Number Theory. 2023. pp. 55-63. DOI: 10.1201/9781003318712-4

[14] Patnaik LR. Cryptography and network for Bachelor of Technology

Computer Science and Engineering. Veer Surendra Sai University of Technology; 2011. pp. 273-303

[15] Gupta A, Walia NK. Cryptography algorithms: A review. International Journal of Engineering and Developmental Research. 2014;**2**(2): 1667-1672. [Online]. Available from: http://citeseerx.ist.psu.edu/viewdoc/download;jsessionid=FEF3E8340DC536679E3C83BF43F1616C?doi=10.1.1.674.7141&rep=rep1&type=pdf

[16] Sagar V, Kumar K. A symmetric key cryptography using genetic algorithm and error back propagation neural network. In: 2015 2nd International Conference on Computing for Sustainable Global Development (INDIACom), New Delhi, India. IEEEExplore; 2015. pp. 1386-1391

[17] Triandi B, Ekadiansyah E, Puspasari R, Iwan LT, Rahmad F. Improve security algorithm cryptography Vigenere Cipher using chaos functions. In: 2018 6th Int. Conf. Cyber IT Serv. Manag. CITSM 2018, no. Citsm. 2019. pp. 1-5. DOI: 10.1109/CITSM.2018.8674376

[18] Ogiela L, Ogiela MR, Ogiela U. Cognitive information systems in secure information management and personalized cryptography. In: 2014 Joint 7th International Conference on Soft Computing and Intelligent Systems (SCIS) and 15th International Symposium on Advanced Intelligent Systems (ISIS), Kitakyushu, Japan. IEEEExplore; 2014. pp. 1152-1157. DOI: 10.1109/SCIS-ISIS.2014.7044798

[19] Submitted to Cs530: Advanced Algorithm Design and Analysis. 2015

[20] Suzaki T, Minematsu K, Morioka S, Kobayashi E. TWINE: A lightweight, versatile block cipher. In: ECRYPT Workshop on Lightweight Cryptography. nec.com; 2011

[21] Das SB, Mishra SK, Sahu AK. Cryptography algorithm. In: A New Modif. Version Stand. RSA Cryptogr. Algorithm. Vol. 767. 2020. pp. 281-287

[22] Rahmani MKI. Cryptographic algorithms and protocols. In: A Step Towar. Soc. 5.0 Res. Innov. Dev. Cloud-Based Comput. Technol., 2021. 2021. pp. 11-20. DOI: 10.1201/9781003138037-2

[23] Rachmawati D, Budiman MA, Wardhono MI. Hybrid cryptosystem for image security by using Hill Cipher 4x4 and ElGamal Elliptic curve algorithm. In: 2018 IEEE Int. Conf. Commun. Networks Satell. Comnetsat 2018 – Proc. 2018. pp. 49-54. DOI: 10.1109/COMNETSAT.2018.8684121

[24] Jangid RK, Mohmmad N, Didel A, Taterh S. Hybrid approach of image encryption using DNA cryptography and TF Hill Cipher Algorithm. In: Int. Conf. Commun. Signal Process. ICCSP 2014 – Proc. 2014. pp. 934-938. DOI: 10.1109/ICCSP.2014.6949981

[25] Semwal P, Sharma MK. Comparative study of different cryptographic algorithms for data security in cloud computing. In: Proc. – 2017 3rd Int. Conf. Adv. Comput. Commun. Autom. (Fall), ICACCA 2017. IEEEExplore; 2018. pp. 1-7. DOI: 10.1109/ICACCAF.2017.8344738

Chapter 4

Pattern Devoid Cryptography

Gideon Samid

Making "Brute Force" the sole attack strategy – And Defending Against it
Hidden Math threatens cyber security; Randomness is the answer

Abstract

Pattern-loaded ciphers are at risk of being compromised by exploiting deeper patterns discovered first by the attacker. This reality offers a built-in advantage to prime cryptanalysis institutions. On the flip side, the risk of hidden math and faster computing undermines confidence in the prevailing cipher products. To avoid this risk one would resort to building security on the premise of lavish quantities of randomness. Gilbert S. Vernam did it in 1917. Using modern technology, the same idea of randomness-based security can be implemented without the inconvenience associated with the old Vernam cipher. These are Trans Vernam Ciphers that project security through a pattern-devoid cipher. Having no pattern to lean on, there is no pattern to crack. The attacker faces (i) a properly randomized shared cryptographic key combined with (ii) unilateral randomness, originated ad-hoc by the transmitter without pre-coordination with the recipient. The unlimited unilateral randomness together with the shared key randomness is set to project as much security as desired up to and including Vernam levels. Assorted Trans Vernam ciphers (TVC) are categorized and reviewed, presenting a cogent message in favor of a cryptographic pathway where transmitted secrets are credibly secured against attackers with faster computers and better mathematicians.

Keywords: randomness, pattern, complexity, mathematical secrecy, user-centric cryptography, subliminal message

1. Introduction

The group think of modern cryptography is that cipher builders are better mathematicians than cipher crackers, hence if the former does not see a mathematical cryptanalytic pathway, neither would the latter. The fact that Alan Turing proved them wrong 80 years ago makes no difference, that is the power of groupthink [1]. Following the revelations of Edward Snowden though, more people suspect that the cryptographic powerhouses are using unpublished math to compromise the security of their targets [2]. A new line of thought emerges: building ciphers that are not based on pattern-loaded algorithms, but rather on the opposite: pattern-devoid algorithms [3–39].

The search for pattern is never exhaustive; hidden layers loom, and are fertile grounds for attackers aiming at mathematics-reliant ciphers. Randomness, on the other hand, is by definition the absence of pattern. Pattern may be viewed as the holes, folds, and protrusions on a mountain you climb, you hang on to them on your way up. Randomness is a perfectly smooth wall, there is nothing to hook up to.

Cryptographers though, are mathematicians in heart, pattern is their thing. They loathe and dismiss the plain logic that argues: pattern-reliant ciphers are threatened by deeper pattern missed by the cipher builder and spotted by the cipher attacker.

The more pattern you detect, the more pattern must be suspected—fertile ground for attackers. It is hard for a defendant to credibly appraise the mathematical insight of his attacker. It is much more feasible to estimate adversarial computing power. This leads to a cipher construction strategy wherein mathematical prowess will be dethroned as the main cryptanalytic tool, and only computing power—brute force—will be left. Such ciphers do not necessarily have to be as perfectly secure as the illustrious Vernam cipher. Their risk of compromise though, must be credibly appraised in terms of the required computing workload. Such appraisal, together with a credible estimate of the attacker's computing power, will generate a good overall estimate of security over time. We, therefore, are off on our way to search for ciphers that will compel their attacker to resort to brute force cryptanalysis, finding mathematical superiority useless. We focus first on symmetric ciphers.

Mathematical cryptanalysis works its way backward. The ciphertext, C, is examined and studied, to reveal its generating plaintext, P, and its operational key, K: C = E (P,K). The strength of a cipher depends on the mathematical properties of the encryption algorithm E, and on the randomness load of the key, K. The stronger E is, the smaller (the weaker) K can be. (E.g.: ECC keys are smaller than RSA keys because ECC is regarded as mathematically more robust than RSA).

We are on a hunt for a cipher that will not rely on the mathematical properties of E for its security (because those properties are vulnerable to hidden mathematical insight) but rather rely on the randomness facing the attacker.

We do not have to look far, 104 years ago, Gilbert S. Vernam patented his now famous "Vernam Cipher" [40] where security is 100% based on the randomness of the key. The Vernam cipher is not based on any hackable mathematical properties. Its security relies on the purity of the randomness of the key. And as is well known, Vernam security is perfect, as has been proven some 25 years later by Claude Shannon [41].

The price paid for this high security was a key as large as the plaintext. This represented too much inconvenience at the time, and hence cryptographers steered away and took the technology of secrets in the opposite direction: using small keys combined with mathematical complexity. This is where we stand today. This trend has developed so much momentum that sidekicks suggesting an alternative, are all but ignored.

Initial application of the emerging Artificial Intelligence Assisted Innovation (AIAI) system [42, 43] shows how fertile it is to reinspect innovative forks of the road, and revisit the path not taken. This is exactly what was done with respect to the old Vernam road junction.

Let R be the randomness used by a cipher to achieve its aim. Let M be the mathematical complexity used by a cipher to achieve its aim (clearly a nebulous definition). Our premise is that M is inherently not a reliable secret-protective source because the greater the mathematical complexity of M, the greater the chance for a lurking mathematical breach strategy, which will be spotted by the attacker, not by the builder of the cipher. R, on the other hand, is 100% reliable, to the extent that R is perfectly random. The greater R, the greater the security. If $|R| \geq |P|$, P—the encrypted plaintext, then the cipher may admit perfect security.

Gilbert S. Vernam set $|R| = |K|$, [40]. Vernam's randomness was captured in the shared key between the transmitter and the recipient of a message. But this is not a necessary requirement. The transmitter may use a priori unshared, unilateral randomness, U, to achieve: $|R| \leq |K| + |U|$

What is needed for this idea to fly are two things:

1. The attacker should not be able to distinguish between K and U.

2. The recipient should not be confused by the impact of U.

If these two conditions (1,2) are fulfilled then the communicators will be able to achieve any desired security, with a limited size key, K, together with a sufficiently large U.

If we set: $|R| = |P| - |K| = |C| - |K|$.

We achieve Vernam security in apparent violation of Claude Shannon's dictate [44–46]: $|K| = |P|$.

Shannon's proof of mathematical secrecy was based on comparing an attacker holding the ciphertext, C, to an attacker that only knows the size of the ciphertext [41]. If the two attackers face the same challenge then there is no advantage to having knowledge of C over having only knowledge of the size of C—which is what Shannon defined as perfect security.

If the C-knowing attacker has to check out $|K|$ options, and the *C-not-knowing* attacker has to check $|P|$ options then if $|K| = |P|$, there is really no advantage to knowing C. However if $|K| < |P|$, then the C-knowing attacker has an advantage over the C *not-knowing attacker*.

Three observations: (i) If $|K| < |P|$ but still very large, then security is not "perfect" but may still remain formidable. One could seek to construct ciphers wherein the security deterioration as the used plaintext exceeds the protective randomness, is happening very slowly; (ii) what if $|K|$ is not known to the attacker? Or say, if $|K|$ is unbound? In that case, the attacker cannot conclusively end the search for the key space.

The third observation is the crux of this treatise. Vernam is a cipher where security receives no contribution from mathematical complexity, M = 0; its security is generated only by the randomness of the key. We can therefore define a class of ciphers, to be called "Trans Vernam" which have this very property: their security is not generated from mathematical complexity but solely from the amount of randomness used. The more randomness—the better security, for a key larger or equal to the encrypted plaintext, this security is perfect.

An attacker of a Trans Vernam cipher is cornered to use a Brute Force attack only, and hence the cryptanalytic burden ahead is represented by the quantity of randomness, R, facing the attacker. Vernam generated the required R through the shared key, K, but this is not a requirement. R can be generated from a shared key K and unshared randomness U. U will be randomness generated by the transmitter without pre-coordinating with the recipient.

The attacker of the Trans Vernam Cipher facing R, (of unknown size) will have no information as to how R divides to K and U. Perhaps U is zero, and R = K? Therefore the brute force attacker will have to check every R option, counting $2^{|R|}$ possibilities (again, $|R|$ is not known to the attacker). The security of the ciphertext is determined first by the transmitter and the recipient together, setting up the key, K, and then by the unilateral determination of the transmitter as to the value of U. The transmitter indeed is the best party to decide on how much security a ciphertext deserves, and hence how much U to use, since the transmitter knows how sensitive the transmitted message is. The transmitter determines the value of R, and hence controls the security projected from the unleashed ciphertext.

It is worth emphasizing that the secrecy regarding the size of the key is critical for the Trans Vernam Strategy to work. If the key size is known, and everything else is known to the attacker (Kerckhoffs' principle, [47, 48]) except the contents of R, then the attacker could train their brute force on K and void the impact of U.

The larger the size of the key, K, relative to U, the greater the chance that a smaller key will offer a plausible (and false) decryption of the ciphertext. And hence, an attacker, finding a reasonable key, is inherently plagued by the doubt of it being a false key, steering the attacker to a misleading decryption of the ciphertext.

For this scheme to work, it is necessary for the recipient to be able to use their knowledge of K to decrypt C to P without being confused by the impact of U. There are several established ciphers that accomplish this. These are Trans Vernam Ciphers.

Sources: [25, 31, 49, 50].

2. Characterization of Trans-Vernam Ciphers

Two shared characteristics: (i) the size of the key is part of its secret, (ii) the Trans Vernam key is reusable.

Third differentiating characteristics: (iii) ciphertext absorbing, or not absorbing the unshared randomness.

With the size of the key being part of its secret, the brute force attacker will never be able to conclusively terminate their search, even if a good key candidate was found. It may be a mistake, and the right key is a larger one. This uncertainty can be used by the communicators by sending meaningless random bits between them. Their attackers will exhaust themselves looking for ever higher keys to crack the communication.

The Vernam key is not reusable. New key bits must be furnished to encrypt new plaintext. This generates a daunting synchronization challenge which is a basic reason for the unattractiveness of Vernam. However, Claude Shannon's proof does not require non-reusability, it only requires key size. Trans Vernam Ciphers operate with a large enough key, which is being used over and over again.

To wit: Using Vernam, if the overall plaintext P is divided into smaller sections $P_1, P_2, \ldots, P_t$, then the respective Vernam key K, will have to be divided into same size keys: $K_1, K_2, \ldots, K_t$, and encryption proceeds as: $C_i = E_{Vernam}(P_i, K_i) \ldots \text{for } i = 1, 2, \ldots, t$.

In a Trans Vernam cipher one proceeds as follows: $C_i = E_{Trans-Vernam}(P_i, K) \ldots \text{for } i = 1, 2, \ldots, t$.

So no synchronization is needed, and more importantly, a Trans Vernam cipher may conveniently be used by a multiplicity of communicators without requiring all parties to follow all communications among other parties, as the case is with Vernam.

Some Trans Vernam Ciphers (TVC) generate the required randomness through a sufficiently large key, and some use unshared randomness, U, with small more manageable keys. The use of U may be reflected in a ciphertext larger than its generating plaintext $|P| < |C|$. The cipher is designed such that the recipient can readily ignore the inflated part of the ciphertext, and credibly extract P from C, using K.

When a size-preserving cipher, E, relies on a key K which is smaller than the message P, then it means that out of $2^{|P|}$ possible plaintexts of size |P| bits only $2^{|K|}$ are viable. The rest are not. The reduction from $2^{|P|}$ possibilities to $2^{|K|}$ possibilities is affected by the pattern inherent in E. This pattern is in the cross-hair of the non-brute force attacker. To the extent that $|K| \rightarrow |P|$ that is the extent that pattern shrinks, and security shifts to randomness. When |P| spills over |K| the security of the cipher deteriorates. However, the rate of deterioration can be credibly appraised by the cipher users, who will decide at each instant if the risk-benefit balance will make it worthwhile to continue using the same key, or arrange for a replacement. It is a main objective for the Trans Vernam Cipher designer, to construct a cipher where said deterioration is as slow as possible [14, 51].

Ahead we discuss the two main categories of TVC: (i) ciphertext-inflated TVC, and (ii) ciphertext-not-inflated TVC, followed by means to handle the extra burden of an inflated ciphertext.

3. Uninflated ciphertext TVC

Identifying three ciphers of the ciphertext uninflated category. One is based on transposition, another on a multi-dimensional roadmap taking the role of the shared key. The third is based on a scheme by which the communicators will iteratively replace an existing key with a new key without this replacement being compromised by the attacker.

3.1 Complete transposition cipher

The following is a short overview of the cipher, defined in "Equivoe-T: Transposition Equivocation Cryptography." [52], "The Ultimate Transposition Cipher (UTC)." [53, 54], and in "Equivoe-T: Transposition Equivocation Cryptography" US Patent 10,608,814 [55].

A plaintext P of size n = |P| bits will have $T = n! / (n_0! * n_1!)$ permutations, where n_0 and n_1 are the number of 0 and 1 in P, respectively: $n_o + n_1 = n$. The highest value of T is $T_{max} = n! / 2\,(0.5n)!$. By using a key such that: $K_{max} \geq T_{max}$

one achieves complete permutation equivocation.

It is easy to beef this security up to full Vernam by constructing a string Q as: Q = P⊕"11… .1"

and concatenating: $\pi = Q \,||\, P$.

π has 2n bits, n of them are 0 and n are one. It has $T = (2n)! / 2n!$ permutations, hence a key space where: $K_{max} > (2n!)/2n!$

will elevate the security of P to Vernam grade. But unlike Vernam the transposition key K_T does not need synchronization and it can be used for plaintexts smaller than itself. What is more, if |P| grows larger than |K| then the security level drops down from perfection, but this drop-down happens very slowly so that high security is maintained.

The complete transposition cipher uses a transposition algorithm with a distinct advantage. Most transposition algorithms are size defined. Namely a simple mapping list will dictate how n bits will be reshuffled around to different places, but these reshuffling instructions specify a particular value for n. The complete transposition cipher is defined over any value of n. It defines a sequence of shipping bits from P to another list, P^t of the same size where the order by which the bits are picked for shipping is determined by the value of the key. The algorithm is symmetric and P^t will be readily returned into P.

Clearly, like with Vernam, the complete transposition cipher has M = 0. Its security is not based on any mathematical complexity. It is based solely on the fact the transposition key K^t is randomly selected from a large enough key space.

It has been proven that the space for a single integer key, K, operating through the complete transposition algorithm will cover all the permutations. As described above if P is separated into parts P_1, P_2, ... P_t then the same transposition key, K will transpose each P_i to P_i^T over its bit size $|P_i|$ without the need to synchronize.

3.2 Space Flip Cipher

This is a brief overview of the cipher described in detail in "SpaceFlip: Unbound Geometry Cryptography" [56, 57], "SpaceFlip: Unbound Geometry Security" US Patent 10,790,977 [58].

This cipher fashions a key in the form of dimensionality-undetermined space, S, comprising a distance geometry [59]. Namely, each of the points of the space has a random distance from any other point. This space is clearly not a metric space, and does not obey proximity laws. The points comprising the space are letters of an alphabet. A line on this space is defined as a sequence of points where the next point in the sequence is constrained by the points that make up the line so far. The determination of the next point is dependent on all the distances from this point to all the points not already on the line. Every point has a next point until all the points are part of the line. Thereby each line can be regarded as a permutation of the points in S.

To send a plaintext letter A, the transmitter may randomly choose a letter B, and mark a line from it. This line will encounter the letter A after the s steps. Therefore the combination {B,s} will be interpreted as the letter A, by the recipient who is working with the same space S. Next time when A is to be sent out as a plaintext letter the transmitter may choose another letter, say, D, and mark a line off it. This line will encounter the letter A after s' steps, so the combination {D, s'} will be interpreted as A by the recipient aware of S.

By choosing the alphabet large enough, the security will be robust enough—again only through randomness, no mathematical complexity. The 0.5 t(t-1) distances marked on S comprising t points are randomly chosen and so is each ciphertext letter. By choosing as alphabet all the possible p bits long strings (an alphabet comprising $2^p = t$ letters) the communicators determine the size of S and the level of the projected security.

To be accurate the ciphertext for this cipher is roughly twice as large as the plaintext, but this should be considered a moderate increase. This cipher, SpaceFlip, can also be implemented with size preservation, only with less security. The transmitter will randomly determine a step count value, s, and then communicate every plaintext letter A with the letter A' such that A appears as the s point on the line in S that begins with A'. This can be run t times without repetition, making it for that measure equivalent to Vernam, without the inconvenience of Vernam.

We will discuss in the next section how SpaceFlip can be implemented with a size increase option.

3.3 Forever Key Cryptography

This is a brief overview of the cipher described in detail in "FAMILY KEY CRYPTOGRAPHY: Interchangeable Symmetric Keys; a Different Cryptographic Paradigm" [60], "SpaceFlip Plus: Ordinal Cryptography" US Patent 11,159,317 [61].

This cipher is constructed to allow for the infinite number of keys to be interchangeable, namely have the same effect. Each of these so-called "family of keys" K_1, K_2, ... , K_i, will encrypt a given plaintext to the same given ciphertext. On its face it seems counterproductive: an attacker will now have an infinite number of keys to hunt, any one of those keys will compromise the cipher. Why make it easier on the attacker?

The answer is plain. An attacker that would successfully compromise a transmission and will extract the plaintext from the ciphertext is likely to have spotted some key K_a which is different from key K_u, which the users have used. The most that an attacker can accomplish is to figure out the entire family of interchangeable keys. There is no way for the attacker to nail down which of the infinite number of keys was actually used by the users. The ciphertext simply does not contain any information that points to the particular key that was used to generate it. That is the power of interchangeable keys; they conceal the identity of the key that was actually used.

This advantage that the users hold over their attacker can be used to exchange a transformation formula, to compute a derived key K'_u from the used K_u, and continue their secret communication with their new-shared key, K'_u. The key derivation formula is constructed such that every input will generate a different output. And since the attacker does know the identity of K_u, they would not be aware of the identity of K'_u either, although the transformation formula is exchanged in the open.

The users can then continue their communication using the derived key K'_u', while dismissing K_u. From the point of view of the attacker, this will be as if the users agreed on a new key in secret. The attacker will not be able to exploit any information garnered from cryptanalyzing K_u to learn anything about K'_u.

After some use the communicators will repeat this operation and derive a third key, K''_u, and so on. Even if the attacker cracks the communication that uses some key K^*_u, the identity of K^*_u is not extracted, and hence when K^*_u is being transformed to $K^{*\prime}_u$, the identity of the new key is not known either, so the users operate as if they started to communicate with their original key, K_u. In other words, this is a mechanism to keep a finite key for indefinite use.

In practice, the number of keys to be considered by the attacker is not infinite because there is a practical limit to how large such a key can be. But this implies that the users can approach this infinite protection by choosing keys that are larger. Since the only way to crack this cipher is by using brute force, the users can credibly estimate how much plaintext they can safely use before their family of keys will be flashed out by the attacker. Based on this estimate the users will switch to the next key before that measure of plaintext is processed.

Because in practice the keys in the family of keys are of a finite count, then the security of this "forever key" cipher is not infinite either. Albeit, its level of deterioration can be credibly appraised. And when needed the transformation of the keys will build very large keys, so that security is upheld.

4. Inflated ciphertext TVC

In this category, ciphers pack the unilateral randomness injected by the transmitter into a containing ciphertext that hence is getting bigger. This is the price paid for security that is not vulnerable to hidden math. As long as the recipient can readily shake off the extra ciphertext material, the only inconvenience with these ciphers is the much larger file to be communicated as ciphertext (no need to store the inflated ciphertext). With today's technology, this is not a big burden even for large documents. When it comes to images, audio, and video files such a large size multiplier for the ciphertext does present a problem. We will see ahead how to navigate this hurdle.

We review here the following ciphers: (i) SpaceFlip, (ii) BitMap, (iii) BitFlip, (iv) The Unary Cipher.

4.1 Increased Ciphertext SpaceFlip

This is a brief overview of the cipher described in detail in "SpaceFlip: Unbound Geometry Cryptography" [56, 57], "SpaceFlip: Unbound Geometry Security" US Patent 10,790,977 [58].

SpaceFlip as described above (in Section 3) can be deployed in a ciphertext-increased mode. The procedure is as follows. To transmit the letter A to the recipient the transmitter will randomly choose any letter from the alphabet A', and then randomly choose an integer g from an arbitrary g-space from 1 to g_{max}. Next, the transmitter will randomly choose (g-1) integers in the range $1 \leq r_i \leq t$, for i = 1, 2, ... (g-1) where t is the number of points in S.

Next, the transmitter will mark a line L_1 from A' to the letter r_1 steps ahead, say, letter A". From letter A" the transmitter will mark a line comprised of r_2 steps, ending up with letter A"', from there to the next letter r_3 steps away. Repeating the same sequence for all the (g-1) distances, the transmitter will end up at some letter B. The line from B will encounter letter A r_g steps ahead. The transmitter will now send over to the recipient the letter A' and the g distance integers: r_1, r_2, ..., r_g.

Marking this information on the shared space S, the recipient will spot the letter A very readily.

There is no limit to the value of g. Large g values lead to large ciphertexts. As the key is being used more and more, the transmitter uses more and more unshared randomness, picking larger and larger g values.

Here too, no pattern, the distances between the t points on S are fully randomized. There is no math to crack, brute force is the only viable attack strategy, and hence the size of S is the sole source of security. The users can stop using a given space S (a key) when the amount of plaintext used through it is exceeding a security threshold. This SpaceFlip protocol will achieve Vernam security with the convenience of a Trans-Vernam cipher.

4.2 BitMap

This is a brief overview of the cipher described in detail in "At-Will Intractability Up to Plaintext Equivocation Achieved via a Cryptographic Key Made As Small, or As Large As Desired - Without Computational Penalty." [62], "BitMap Lattice: A Cyber

Tool Comprised of Geometric Construction", US Patent 10,911,215, [63]; "Denial Cryptography Based on Graph Theory." Gideon Samid (2004) US Patent 6,823,068 [64].

This cipher is a map where the points are associated with letters of a particular alphabet. Points are connected to their direct neighboring points but not to other points. The connections themselves may be viewed as "walking bridges" allowing a traveler to walk from one point to the other. Every bridge is marked by a letter of the same or of a different alphabet that marks the points on the map. The basic idea of the cipher is that the plaintext is viewed as a travel guide. Each successive letter in the plaintext indicates the next travel destination. A plaintext comprising p letters will then be associated with a travel path on this map, comprising p visited spots. And since the passage from spot to spot requires passing through a bridge, and each bridge is marked with a letter, then the same pathway that was identified by the plaintext, is similarly identified by listing the successive bridges traversed by the traveler. It is a simple principle: a path on a map can be described by a list of successive destinations, or equivalently by a list of successive bridges one walks on. The plaintext is seen as a travel guide pointing to the visited destinations; the ciphertext, by contrast, is seen as the list of bridges one passes through when taking the same path. Each bridge is associated with a letter, so the pathway described as crossed bridges will manifest itself as a series of letters—the ciphertext.

The cipher is designed so that any possible plaintext can be mapped into a pathway, and every possible pathway can equally be described by a list of crossed bridges. The transmitter will mark the points (spots) on the pathway corresponding to the plaintext, then describe the same pathway by marking the crossed bridges, and when the list of crossed bridges is assembled, it is communicated to the recipient.

The recipient on his part will use the ciphertext to mark the same pathway on their copy of the map. Once marked the recipient will read out the visited spots and mark the letters represented by these spots in a sequence—thereby reconstructing the plaintext.

The attacker without possession of the map (the key) will not be able to reverse the ciphertext into the plaintext. The configuration of the map is randomized; the marking of the points on the map and the markings on the bridges are all highly randomized, subscribing only to a weak restriction. Thereby the map projects no analytic complexity. It is fair to say that only brute force has a prayer and a hope to crack this cipher.

It is worth noting that the attacker does not know how big the map is. Each spot and each bridge can be visited and crossed countless times. So a very small map will encrypt and decrypt a very long message if necessary without betraying the size of its key.

Building a large key is providing share randomness, K. Albeit, BitMap will allow a transmitter to inject unshared randomness as follows: the plaintext alphabet A' is deemed to be comprised of $l - 1$ letters. Another letter, letter number l is then added by injecting it between any two successive letters that are identical (in the plaintext). This operation removes all instances where two instances of same letter are written one after the other. The transmitter can now replace any letter in the l letters alphabet, A, of the plaintext, with any number of identical letters next to each other. This inflates the plaintext to any desired size. Doing so will in turn lead to an inflated pathway and an inflated ciphertext. The recipient though, recovering the long plaintext, will simply shrink all letter repetition to a single letter and thereby extract the original plaintext.

Illustration: the string ABC will become AAAABBBBBBCCCCCC. This inflated plaintext will then mark a much longer pathway on the map. This longer pathway will be translated to a long ciphertext that would confound the attacker. The intended recipient will extract the inflated ciphertext but will shrink it to size. Every string of consecutive same letter, like AAAA and BBBBBB, will be replaced by a single same letter: AAAABBBBBBCCCCCC → ABC.

The attacker, in possession of the ciphertext, does not see the duplication. It does not show on the bridge-list. As a result, the attacker wrestles with a long ciphertext, not knowing which parts, if any, are the real concealed message and which parts are confounding noise.

4.3 BitFlip

This is a brief overview of the cipher described in detail in "BitFlip: A Randomness Rich Cipher" [65]; "BitFlip Cyber Demonstration" [66]; "Transmitter for Encoding Information with Randomly Flipped Bits and Transmitting That Information Through a Communication Channel" US Patent 10,728,028, [67]; "Advanced BitFlip: Threat Adjusted, Quantum Ready, Battery Friendly, Application Rich Cipher" US Patent 10,541,808, [68].

The idea behind BitFlip is to have a large number of ciphertext letters map into a single plaintext letter in parallel to having a large number of ciphertext letters map into no plaintext letter. The first attribute resists pattern recognition through the randomized selection of ciphertext letters among the many that map into a given plaintext letter. The latter attribute allows the user to freely inflate the ciphertext with false letters, which the intended reader will readily recognize as such, while the attacker will have to regard them as message-bearing. By carefully subjecting all choices to randomization, the users expunge any pattern from the construction of the cipher, cornering the attacker to brute force attack strategy—the efficiency of which can be credibly assessed by the BitFlip users.

BitFlip works on some alphabet A comprising l letters. Each letter, L_i ($i = 1, 2, \dots, l$) is represented by a bit string of size n bits—selected randomly. Each letter is associated with "Hamming distance" value, h_i, where $0 \le h_i \le n$. A ciphertext letter c is a bit string of size n. Letter c is decrypted to plaintext letter L_i iff: $H(c, L_i) = h_i$

where $H(c, L_i)$ is the Hamming distance between c and L_i.

Ciphertext letters that don't decrypt to any of the l plaintext letters are discarded.

The selection of the n-bit string representation for each of the l letters is randomized. The selection of the l h_i values is randomized, and the selection of the ciphertext letter that decrypts to a given plaintext letter is randomized. The peppering of the ciphertext with so-called decoy ciphertext letters that are meaningless, is also randomized. There is no thread of pattern to crack here. The attacker is cornered to brute force attack strategy.

4.4 The Unary Cipher

This is a brief overview of the cipher described in detail in "A Unary Cipher with Advantages over the Vernam Cipher" [69]; "Unary Cryptography Demonstration Site" [70]; "Mixed Unary Cryptography" US Patent Application 17/323,908.

A bit string x indicating an integer of value v, can be expressed through a string of (v + 1) "0", concatenated to a list of (r + 1) "1", where r is the number of leading zeros in x. This, so-called unary expression is larger in bit count, but it is otherwise equivalent to the original expression. The two can be derived one from the other.

This property can be used as follows: a plaintext P is randomly broken down to n consecutive substrings: $P_1, P_2, \dots, P_n$. The value of n, and the size of the substrings $|P_i|$ for i = 1, 2, ..., n is randomly selected, as long as: $|P| = \Sigma\, |P_i| \dots \dots$ for $i = 1, 2, \dots, n$

Each P_i is then mapped to its corresponding unary expression, thereby writing P in a combined unary fashion, P^*. P^* may be peppered with "0,1" element because these elements vanish when transposed back from unary format.

P^* is then transformed with a complete transposition key as discussed above: $P^* \rightarrow P^{*T}$.

P^{*T} is the ciphertext. $P^{*T} = C$.

It can be shown that any value of P' from some high-level $Q > P$ to zero may be encrypted to the same ciphertext C, thereby projecting a functional equivalent with Vernam. Its advantage over Vernam is that the key (the transposition key) can be used over and over again, and no synchronization is needed.

P^* may be wrapped with a header of the form 00....1 and a trailer of the form 11.....0. These are two more options to pad the ciphertext in a way that would not confuse the intended reader, but will build a growing cryptanalytic burden.

To further increase the cryptanalytic burden one will prepare the pre-transposition string as one with equal count of zeros and ones, as follows: (i) compute $P^\wedge = P \oplus$ 11......1 (ii) concatenate P with $P^\wedge$: $P^{**} = P^* \,||\, P^\wedge$, then transpose P^{**} (of size $2|P|$).

To the extent that the transposition operation is complete, this cipher projects Vernam grade security over its secret size key.

5. Split Security Solutions

The price paid for randomness-based security is a large key, and in many cases a large ciphertext. In order to keep projecting high security from the same key, k, the ciphertext will have to be longer and longer than the plaintext. This increased ciphertext length is looming to become a more and more serious problem for large plaintexts. When the materials to be encrypted are audio files, images, or video files then the much larger ciphertext may be prohibitive. This challenge can be handled via Entropic Impact Discrimination.

5.1 Entropic Impact Discrimination

This is a brief overview of the cipher described in detail in "Split Security Solutions", US Patent Application 17/510,324 [71].

A plaintext P may be divided into meaning-bearing elements $m_1, m_2, \dots, m_t$. Each meaning-bearing element is associated with an entropic impact that reflects the advantage gained by an adversary in case this element is exposed. The elements may then be divided into low-impact elements which have an entropic impact below a given threshold and high-impact elements which have an entropic impact above that threshold. The latter is slated to be encrypted with a Trans Vernam cipher, and the former are encrypted with a size preserving cipher. Thereby the users decide how much inconvenience (large ciphertext) to put up with, in order to get a given level of security.

The high-impact selection may be automatic. For example, facial recognition software will identify faces in an image or in a video, and mark these faces for TVC encryption. The rest will be encrypted with size-preserving ciphers. In the worst-case scenario, the adversary will see what the image shows but will not figure out who the people in the image are.

6. Spontaneous Cryptography

In the 1970s, cryptographic science made a dramatic leap ahead. It enabled two strangers to practice cryptographic protocols despite having no prior shared key to rely on. Spontaneous cryptography, or cryptography between strangers, revolutionized the practice of commerce; it became a key enabler of our migration toward cyberspace. Alas, the prevailing schemes are heavily reliant on mathematical pattern, which very likely hides deeper patterns with which to crack this cryptography. So in order to maintain the great benefits introduced by spontaneous cryptography it is suggestive to investigate constructing bilateral secrecy on randomness.

Here too, a scroll back to the pages of history proves helpful. Much as Vernam revisited helps us with ordinary encryption, so Ralph Merkle is a new pointer for spontaneous encryption. Unlike his followers Diffie and Hellman, Ralph Merkle based his original idea for strangers developing a bilateral secret while exposed on the network, not on mathematical complexity but rather on a temporary advantage claimed by two communicators, who can solve a riddle a bit faster than their attacker. The communicators then use this temporary secret to secure a permanent secret. Merkle's idea was for one communicator to present the other a list of difficult computational tasks, for which the submitter already knew the answers. The recipient chooses randomly one of the various computational tasks, computes it, and communicates the result to the sender. The value of the answer tells the transmitter which one of the tasks the recipient chose, and that information qualifies as a temporary secret until the attacker will compute the entire list and also find out which task the recipient chose.

An advanced version of Merkle's randomness (not pattern) based spontaneous cryptography is offered by FigLeaf.

6.1 FigLeaf

This is a brief overview of the cipher described in detail in "Randomized Bilateral Trust (RABIT): Trust Building Connectivity for Cyber Space (FigLeaf)" U. S. Patent 10,798,065, [72].

The FigLeaf idea is based on the familiar "birthday paradox": it turns out that a group of only 23 people have a 50% chance to include two people with the same birthday, month, and day of the month. Similarly, two strangers would agree to randomly pick n mathematical items from a large list L of such items. The value of n can be adjusted to make it x% likely for the two item selectors to have picked the same item (a picked item remains in L). The two then start a dialogue. The selected items have various properties. One communicating stranger randomly selects a property and tells the other the n values of this property in the n items she selected. The other communicator can then exclude any of his selections for which the value of this property is not on the list submitted by the first communicator. His list shrinks n → n'. Next, the list-recipient selects another property, and hands over the list of its n' values. The first communicator will delete from his list all the items that have a value for that property that is not in the submitted list. This will shrink her list n → n". By repeating this protocol the two strangers would either realize that they have no item in common, and in that case they will restart the protocol, or they would both identify the one item they both randomly picked. It will take an outside observer much longer time to use the information exposed in the protocol to spot the shared item. Any of the unused properties of the shared item will qualify as a temporary secret, which the communicators will use to secure a permanent secret key if necessary. For cases like money transfer the temporary secret will do. Once the money is transferred the shared secret becomes useless.

Unlike the Diffie-Hellman scheme, FigLeaf is randomness based, and the only route of attack is brute force.

7. Extended applications

The power of cryptography to hide a secret in an exposed capsule (a ciphertext) serves a variety of applications beyond enabling a conversation in a hostile environment. Most common among them are (i) cryptographic authentication, applicable to humans and things alike, (ii) pattern concealment, and (iii) graded randomness applications. To the extent that mathematical cryptography is vulnerable to the original application, it is similarly vulnerable to its extended applications. And to the extent that Trans Vernam Ciphers cure the vulnerability of nominal cryptography for secret communication, it similarly cures the vulnerability presented itself when used to authenticate a document, a person, a thing.

Presenting (i) authentication applications, (ii) pattern concealment.

7.1 Authentication

Authentication is a process where a Verifier verifies that another party, "The Prover", is in possession of a piece of information, P. The challenge is repetition: to ensure that the proof does not disclose to an attacker how to falsely claim bona fide possession of P. Hence, direct exposure of P is not an option. There are three prevailing ways to hide P: (i) pass P within a secure channel, (ii) apply private-public key, (iii) apply randomness. The second option is by far the most popular. Yet, it is the first target for quantum cryptanalysis, and its days are numbered. Good alternatives are in order.

Presenting: (i) challenge-response authentication (ii) randomized authentication protection.

7.1.1 Challenge-response authentication

> *The following is a short overview of the cipher, defined in "Efficient Proof of Knowledge of Arbitrarily Large Data Which Remains Undisclosed" US Patent 10,594,480 [73].*

To prove possession of a body of data, P, by a "Prover", the "Verifier" will randomly pick a number R, and a function f, and pass both to the prover (in the open). The prover will use R, and f to transform P into a bit string Q: Q = f(R,P). Next, the prover will parcel out Q to n distinct concatenated substrings $q_1, q_2, \ldots, q_n$. $Q = q_1 \| q_2 \| \ldots \| q_n$.

The breakdown of Q to n substrings will be carried out according to a shared rule which will ensure that: $q_i \neq q_j, \ldots$ for $i \neq j$, and $i, j = 1, 2, \ldots, n$.

The breakdown rule is not secret. The prover will then apply the Complete Transposition Cipher, as described above, to reshuffle the n substrings to a transposed Q: Q^T.

Q^T will be passed to the verifier. The verifier will similarly identify the n substrings $q_1, q_2, \ldots, q_n$, and confirm that Q^T can be constructed from $q_1, q_2, \ldots, q_n$ in some order.

One may note that the verifier may verify the prover being in possession of P, without the verifier being in possession of P. All that the Verifier needs in order to do their job is to have n substrings: $q_1, q_2, \ldots, q_n$, without having knowledge of the particular permutation thereto that assembles them to Q.

Since the transposition is complete, the attacker in possession Q^T will have first to list all the possible ways, m, in which Q^T can be divided to an unknown number of substrings, $n_1, n_2, \ldots, n_m$, compliant with the substrings division rule, then for each possible number of strings consider all the permutations thereto.

By selecting f, and R so as to generate Q of any desired size, the verifier will ensure that the brute force load on the attacker will exceed their ability to extract P in a timely manner. Because the combination of R and f is not repeated, the attacker will have to extract P from Q^T in order to prove possession thereof.

7.1.2 Protecting authentication databases

The following is a short overview of the ciphers, defined in "Method for Inhibiting Mass Credentials Theft" US Patent 10,395,053 [74].

Identity authenticators happen to be organizations serving a large number of customers. Using secure channels these authenticators receive the credentials of their customers, compare them to their records, and thereby authenticate them. The records kept by these organizations are at the cross-hair of sophisticated attackers since a single penetration will net the private data of all the customers of the victim organization. This can be prevented by allowing the authenticators to perform the authentication without keeping their customers' data in their authentication records. It seems impossible at first glance, how would one authenticate unknown data? It can be done through zero knowledge techniques that are math-loaded. Here we present a randomness-based solution.

Information submitted for authentication is written as bit strings. Bit strings are conveniently written in Base64. Base64 is a language comprised of 64 letters. We map the letter number i in this alphabet (i = 1, 2, ..., 64) to a bit string comprising i bits—disregarding the identity of the bits. We now determine the identities of these message-bearing bits through a source of randomness. This will yield a fully randomized layout, R, of the message A submitted for authentication. Let's divide A to n bits segments: $A = a_1 \| a_2 \| \ldots a_m$ where $|a_i| = n$. Each segment will also be randomized. Let us now flip $h < n$ bits in each segment. Such flipping will not affect the prime message, written in Base64, but it will shift the bit expression of the Base64 letters. Each message a_i will be shifted to a'_i, where a_i and a'_i exhibit a Hamming distance h between them: $h = \text{Hamming}(a_i, a'_i)\ldots\ldots \text{for } i = 1, 2, \ldots, m$.

We now can store the shifted expression: $A' = a'_1 \,||\, a'_2 \,||\, \ldots\, a'_m$ in the server's database, and send the customer the message $A = a_1 \,||\, a_2 \,||\, \ldots\, a_m$ to use when authenticating herself.

The prime message (written in Base64) will be the same both for the record kept on the customer's phone and for the record kept in the server's database. When the server receives the message from the customer, the server first authenticates the prime message (account number, name, password, etc.), and then examines the bit identities of the submitted bit string.

If the Hamming distance between the customer's record and the server's record for all sections of A is h, then the transaction goes through. If the test fails, the transaction stops, and a response protocol is activated.

Should a hacker break into the server and copy its records, they will not harvest the customers' records. They will get a hold of the server's data. Should a hacker attempt to steal a customer's identity using her credentials that were stolen from the server, then the server will immediately realize that the Hamming distance between its records and the data submitted for authentication is zero and not h. An alarm will sound to alert the server and conclude that a hacker pretends to be a customer. The server is further alerted to the fact that the server was compromised. Once so realized the server will simply refresh its records, maintain the prime message but change the Hamming distance from h to $h' \neq h$. This simple act will void the hacker's harvest. The stolen data which has a Hamming distance h from the customer's record will not enable the hacker to contrive credentials that would exhibit a Hamming distance h' from the servers' records. That is because the hacker does not know which bits were flipped and which were not—this choice was made randomly. The net effect of this tool is that (i) a breach is instantly discovered, and (ii) is readily recovered from. This recovery is swift and painless without bothering the customer.

7.1.3 Authentication of material items

The following is a short overview of the technology defined in "Proving Material Identity with Quantum Randomness – Financial and General Applications" US Patent 10,754,326. [75], "BitMint Hard Wallet: Digital Payment without Network Communication: No Internet, yet Sustained Payment Regimen between Randomness-Verifiable Hard Wallets" [76].

Counterfeiting material items is an advanced fraud industry, affecting mainly manufactured items of value. Governments are in a race with counterfeiters over banknotes, passports, and various licenses and documents. Manufacturers suffer when the market is flooded with look-alike products that steal their customers and destroy their reputations.

Much as the prevailing cryptography is opting for greater and greater mathematical complexity to fend off cryptanalysis, so do manufacturers, adding hologram signatures and other physical complexities to remain one step ahead of the counterfeiters. We have seen how randomness is an alternative solution strategy against cryptanalysis; the very same principle applies to material authentication: chucking the unified complexity race in favor of randomized complexity.

A pattern-loaded manufacturing complexity will eventually be deciphered. And once so the counterfeiter will flood the market with hard-to-detect counterfeits. Applying randomness, each item has its own unpredictable signature, so there is no one "secret" that works for all items of the same kind. The counterfeiter will have to tailor the counterfeit to individual signatures. What is more, the randomized signature is comprising a very large number of measured properties. Some are published on a public ledger, allowing the verifier to compare the published and the measured. Albeit, many more properties are not published a priori, and only released to the public upon demand. The verifier will check the submitted item against the just released properties—a counterfeit will fail the test.

7.2 Pattern concealment

The following is a short overview of the cipher, defined in "Effective Concealment of Communication Pattern (BitGrey, BitLoop)" US Patent 10,673,822, [8].

In many practical circumstances attackers gain a consequential advantage by analyzing the pattern of communication traffic: who writes to whom, when, how much, how often, etc. While encryption per se hides the content of the communication, it does expose its pattern. Trans-Vernam ciphers can be used to cure this deficiency. TVC may inflate the ciphertext at will, while the intended reader will not be confused by the meaningless bits and properly interpret the meaningful bits, as described herein. This situation may be exploited by establishing a fixed bit rate among the communicators. The bit flow will range from no-messaging, all bits are meaningless, to all-messaging, no bits are meaningless, and any state in between. The communicators will read only the messages intended for them, but attackers will see a steady unchanging bit flow rate, remaining in the dark as to whether anybody talks to anybody, or who talks to whom, how often, and how much.

8. Randomness technology

The most popular source for randomness are algorithms that generate bits sequences that comply with a given (arbitrary) set of rules. John Von Neumann said that anyone generating randomness from algorithms does not understand, neither randomness nor algorithms. Indeed it makes little sense to abolish mathematical complexity with randomness that is itself a product of mathematical complexity. There are two classes of non-algorithmic randomness: (i) physical complexity, (ii) quantum randomness. The former is based on the formidable amount of real-time knowledge that must be processed in order to defeat it, and the latter is based on a first principle of quantum mechanics.

It is noteworthy that perfect randomness can be theorized, not proven. No matter how many tests are conducted over a source of randomness, the results can always be explained as coming from a source where the deviation from perfect randomness is too small to detect in this finite test. This fact casts a thin but present shadow on the assertions for security claimed herein.

Presenting: (i) quantum randomness technology, (ii) physical complexity randomness technology.

8.1 Quantum randomness technology

The following is a short overview of the technology, defined in "Rock of Randomness" US Patent 10,467,522, [77] and in "The Rock of Randomness: A physical oracle for securing data off the digital grid" [78].

Commercial outfits today offer elaborate apparatuses generating quantum-grade randomness [79, 80]. What is needed then is (i) robust packaging, and (ii) effective duplication, (iii) copy protection. These three needs have been satisfied through "The Rock of Randomness". It packs randomness in the chemical structure of the material constituents of the rock. The data, hence, is off the digital grid, which means it is beyond the territory that is subject to digital compromise. One needs to have access to the Rock itself, in order to read its data. The Rock packs very large quantities of data in its molecular composition. It is measured in an analog format, then digitized. Even a small piece of rock may pack an enormous amount of data, beyond what is practical to image in a database.

The Rock is not vulnerable to accidental physical punishment nor to happenstance chemical obstruction. Melting will destroy it, but otherwise it keeps. The manufacturing of the Rock can be duplicated, but given a manufactured Rock it is infeasible to duplicate it. Communicators holding a duplicate of the same Rock each will enjoy the full power of Trans Vernam Ciphers.

8.2 Physical complexity technology

The following is a short overview of the technology, specified in US Patent Application #17063523.

Quantum randomness enjoys the credibility of the most elevated scientists who claim it to be perfect. Only the generators of this randomness are embedded in complex electronics, which are subject to attack. So while the created bit flow is unbiased, the bit sequence that is poured to its consumers may be contaminated. This is one argument in favor of a closer source based on sufficient real-world complexity that is per symmetry tests devoid of any pattern [39, 81]. One such source is a contraption wherein insulating bubbles rise in a conductive liquid, and reduce its effective conductivity. The reduced conductivity suppresses the bubble's flow, (feed-back cycle) which in turn increases the effective conductivity, that now increases the flow of bubbles. The mechanism varies the quantities of the rising gas, as well as its distribution over a range of bubble sizes. The conductivity variance is translated into a randomized bit stream.

This bubbles randomizer will be external to the consuming computers, and be readily replaceable. Unlike a quantum source, this complexity apparatus hinges on a feedback cycle, so that any disturbance will be diffused to high-quality randomness.

A note on the innovation solution protocol: This presentation is a case study for the innovation path known as historic retrace in which one traces the innovative history of a present state, revisits innovative forks of the roads, and examines roads not taken. The idea is that in hindsight the unselected options in that junction point may look more attractive than when first encountered. Pursuing those untried avenues is a choice loaded with possibilities. Progress from natural evolution to science and technology is a zig-zagging path. This thesis emerges from a rigorous practice of the Innovation Solution Protocol (Innovation^{SP}) [42, 43]. It identified the 104 years old Vernam cipher as the fork in the road from where cryptography selected the small-key path, which was the wise choice for the technology of the day. Albeit with the tools we have at present, the road not taken—projecting security through randomness, not through math—is the road that leads to a new cyber vista.

Allegory: The following short tale illustrates the message contained in this chapter. Two pals, Alice and Bob play a game of dice. One throws, the other guesses. A right guess will move 1$ from the dice thrower to the dice guesser. Playing for hours both players end up with just about the same amount of money they brought to the match. This is very disappointing for Alice who is much better educated than Bob and knows everything there is to know about probabilities and computing. So she proposes to Bob to introduce a tiny variation to the game. Instead of throwing one dice, they will each throw two. Instead of guessing in the range 1–6, they will be guessing in the range 2–12. Bob innocently accepts, but no sooner do they switch to two dice than Alice cleans Bob's wallet to his last dollar. While innocent Bob randomly guesses a choice from 2 to 12, Alice uses her probability education and guesses 7 every time. Bob is losing money every night, blaming his bad luck.

One bright day Bob realizes that his losses occurred when the game was switched from one dice to two dice, so he insists on returning to the original mode. Lo and behold Alice's advantage vanishes.

Trans Vernam ciphers—return to Vernam philosophy (not to the Vernam protocol)—are tantamount to innocent Bob returning to the one dice mode where no matter how smart, or how stupid a player is their playing field is level.

9. Outlook

Pattern-Devoid cryptography may turn out to be an important factor in the evolution of human residence in cyberspace. The currency of the digital realm is trust. The digital facade is an effective veil for all sorts of ill-doings and abuse. Privacy is deathly wounded. Today's ciphers operate under the shadow and the suspicion of stealth cryptanalysis. Facing these threats pattern-devoid cryptography offers the credibility of mathematics, and the assurance of the unpredictability of true randomness, now available commercially. Trust will be reconstructed, and life in cyberspace will evolve to its full potential.

Additional information

The first version of this chapter was published as a preprint on the preprint server of the Archive of the International Association for Cryptologic Research, entitled: "Pattern Devoid Cryptography" Nov 20, 2021.

Author details

Gideon Samid
Electrical, Computer and System Engineering, Computer and Data Sciences, Case Western Reserve University, Cleveland, OH, United States

*Address all correspondence to: gideon.samid@case.edu

References

[1] Kahn D. The Code Breakers. London, UK: The MacMillan Co.; 1967

[2] Wallace C. Edward Snowden: The Untold Story. Hamakua District of Hawaii County, Hawaii, United States: Riisas Honaka; 2014. ISBN-13: 9781774858820

[3] Samid G. Anonymity management: A blue print for newfound privacy. In: The Second International Workshop on Information Security Applications (WISA); September 13-14, 2021; South Korea; 2021 (Best Paper Award)

[4] Sharif A, Raihana NI, Samsudin A. Chaos-based cryptography: A brief look into an alternate approach to data security. Journal of Physics Conference Series. 2020;**1566**(1):012110. Available from: https://iopscience.iop.org/article/10.1088/1742-6596/1566/1/012110/meta

[5] Samid G. Cryptography of Things: Cryptography Designed for Low Power, Low Maintenance Nodes in the Internet of Things. 2016. Available from: https://search.proquest.com/openview/8897dc1c4858b327796917b8fcdff7ae/1?pq-origsite=gscholar&cbl=1976348

[6] Samid G. Cyber Passport: Preventing Massive Identity Theft. 2016. Available from: https://eprint.iacr.org/2016/474

[7] Samid G. Drone Target Cryptography. 2016. Available from: https://eprint.iacr.org/2016/499

[8] Effective Concealment of Communication Pattern (BitGrey, Bitloop). US Patent 10,673,822. 2020

[9] Samid G. Encryption Sticks (Randomats). In: ICICS 2001 Third International Conference on Information and Communications Security; 13-16 November, 2001; Xian, China

[10] Samid G. Encryption-On-Demand: Practical and Theoretical Considerations. 2008. Available from: https://citeseerx.ist.psu.edu/viewdoc/download?doi=10.1.1.215.2463&rep=rep1&type=pdf

[11] Samid G. Feeding Cryptographic Protocols with Rich and Reliable Supply of Quantum-Grade Randomness. 2020. Available from: https://eprint.iacr.org/2020/968.pdf

[12] Samid G. Fingerprinting Data. 2018. Available from: https://eprint.iacr.org/2018/503

[13] Samid G. Hush Functions Extended to Any Size Input versus Any Size Output. 2012. Available from: https://eprint.iacr.org/2012/457.pdf

[14] Samid G. Intractability erosion: The everpresent threat for secure communication. In: The 7th World Multi-Conference on Systemics, Cybernetics and Informatics (SCI 2003). Orlando, Florida: Informs, The Institute for Operation Research and the Management Sciences; July 2003

[15] Samid G. Larger Keys, Less Complexity. A Strategic Proposition. 2018. Available from: https://eprint.iacr.org/2018/406.pdf

[16] Samid G. Proposing a Master One-Way Function. 2007. Available from: https://eprint.iacr.org/2007/412

[17] Randomized Bilateral Trust (RABIT): Building Connectivity for Cyber Space. US Patent 10,798,065. 2020

[18] Samid G. Randomness rising—The decisive resource in the emerging cyber

reality. In: Int'l Conf. Foundations of Computer Science | FCS'18 |. Las Vega, Nevada; 2018. Available from: https://www.bitmint.com/RandomnessRising_GSamid_H1o16.pdf

[19] Samid G. Re-dividing complexity between algorithms and keys. In: International Conference on Cryptology in India. Berlin, Heidelberg, New York, London, Paris, Tokyo: Springer; 2001. Available from: https://link.springer.com/chapter/10.1007/3-540-45311-3_31

[20] Samid G. Rivest chaffing and winnowing cryptography elevated into a full-fledged cryptographic strategy. In: Proceedings of the International Conference on e-Learning, e-Business, Enterprise Information Systems, and e-Government (EEE); Athens; 2018. Available from: https://search.proquest.com/openview/8ea94f941732d85fb24512d5e7582820/1?pq-origsite=gscholar&cbl=1976356

[21] Samid G. Shannon Revisited: Considering a More Tractable Expression to Measure and Manage Intractability, Uncertainty, Risk, Ignorance, and Entropy. 2010. Available from: https://arxiv.org/abs/1006.1055

[22] Samid G. T-Proof. 2016. Available from: https://img.chainnews.com/paper/71f69315d015d9fc5dd4ffbc97f87aab.pdf

[23] Samid G. T-Proof: Secure Communication via Non-Algorithmic Randomization. 2016. Available from: https://eprint.iacr.org/2016/474

[24] Samid G. Tailored Key Encryption (TaKE). 2000. Available from: https://eprint.iacr.org/2000/011.pdf

[25] What a 100-year-old Idea can teach us about Cybersecurity. World Economic Forum. 2017. Available from: https://www.weforum.org/agenda/2017/11/what-a-100-year-old-idea-can-teach-us-about-cybersecurity

[26] Samid G. When Encryption is Not Enough–Effective Concealment of Communication Pattern, even Existence (BitGrey, BitLoop). 2019. Available from: https://eprint.iacr.org/2019/556

[27] Downey R, Hirschfeld D. Algorithmic Randomness and Complexity. Victoria Univ. Wellington, New Zealand: School of Mathematics and Computing Sciences; 2010. Available from: http://www-2.dc.uba.ar/materias/azar/bibliografia/Downey2010AlgorithmicRandomness.pdf

[28] Goldwasser M. Probabilistic encryption. Journal of Computer and System Science. 1984;**28**(2):270-299

[29] Hughes R, Nordhold J. Strengthening the Security Foundation of Cryptography with Whitewood's Quantum-powered Entropy Engine. 2016. Available from: http://www.whitewoodencryption.com/wp-content/uploads/2016/02/Strengthening_the_Security_Foundation.pdf

[30] Horváth M. Survey on cryptographic obfuscation. International Association of Cryptology Research. ePrint Archive. 2015. Available from: https://eprint.iacr.org/2015/412

[31] Samid G. The Unending Cyber War. DGS Vitco ISBN 0-9635220-4-3. 2008. Available from: https://www.amazon.com/Unending-Cyberwar-Gideon-Samid/dp/0963522043

[32] Cyber Companion: Attaching a Secondary Message to a Primary One. US Patent 10,541,954. 2020

[33] Live Documentation (LiDO). US Patent 10,733,374. 2020

[34] Samid G. AI Resistant (AIR) Cryptography. IACR Archive. 2023. Available from: https://eprint.iacr.org/2023/524

[35] Samid G. Tesla Cryptography: Powering Up Security with Other Than Mathematical Complexity. IACR Archive. 2023. Available from: https://eprint.iacr.org/2023/803

[36] Samid G. The Prospect of a New Cryptography: Extensive use of non-algorithmic randomness competes with mathematical complexity. IACR Archive. 2023. Available from: https://eprint.iacr.org/2023/383

[37] Blackledge J, Mosola N. Applications of artificial intelligence to cryptography. Transactions on Machine Learning & Artifical Intelligence. 6 June 2020. DOI: 10.14738/tmlai.83.8219

[38] Barrera FY. Understanding Complexity of Cryptographic Algorithms. Bucharest, Romania: Baeldung; 2023. Available from: https://www.baeldung.com/cs/cryptographic-algorithm-complexity

[39] Boddu NG. Problems in quantum tamper-resilient cryptography and quantum communication complexity [PhD dissertation]. Singapore: National University of Singapore (Singapore) ProQuest Dissertations Publishing; 2022. p. 30340228

[40] Vernam GS. Secret Signaling System. US Patent 1310719A. 1918

[41] Shannon's Proof of Vernam Unbreakability. 2016. Available from: https://www.youtube.com/watch?v=cVsLW1WddVI

[42] Samid G. Artificial Intelligence Assisted Innovation. 2020. Available from: https://www.intechopen.com/online-first/artificial-intelligence-assisted-innovation

[43] The Innovation Solution Protocol. (Innovation^{SP}). 2023. Available from: https://InnovationSP.net

[44] Hellman M. An extension of the Shannon theory approach to cryptography. IEEE Transactions on Information Theory. 1977;**23**(3):289-294

[45] Shannon C. Communication Theory of Secrecy Systems. 1949. Available from: http://netlab.cs.ucla.edu/wiki/files/shannon1949.pdf

[46] Niels A. Computability and Randomness. Clarendon, Oxford, UK: The University of Auckland; 2008

[47] Canetti R, Dwork C, Naor M, Ostrovsky R. Deniable Encryption. In: CRYPTO '97Volume 1294 of the Series Lecture Notes in Computer Science. Santa Barbara, California: Springer; 2006. pp. 90-104

[48] Kerckhoffs' Principle. 2023. Available from: http://www.crypto-it.net/eng/theory/kerckhoffs.html

[49] Samid G. The Myth of Invincible Encryption. Digital Transactions Magazine; Security Notes. Chicago, Illinois; Dec 2005

[50] Samid G. User centric cryptography. In: Proceedings of the International Conference on Security and Management (SAM). Las Vegas, Nevada: The 2018 Congress in Computer Science, Computer Engineering, and Applied Computing; 2018. Available from: https://www.proquest.com/openview/a60ecf397b6c46373356a1d4369dce5d/1?pq-origsite=gscholar&cbl=1976342

[51] Samid G. Essential Shannon Security with Keys Smaller than the Encrypted Message the Encrypted Message. 2000. Available from: https://citeseerx.ist.psu.

edu/viewdoc/download?doi=10.1.1.75.1585&rep=rep1&type=pdf

[52] Samid G. Equivoe-T: Transposition equivocation cryptography. International Association of Cryptology Research. ePrint Archive. 2015. Available from: https://eprint.iacr.org/2015/510

[53] Samid G. The Ultimate Transposition Cipher (UTC). 2015. Available from: https://eprint.iacr.org/2015/1033.pdf

[54] Samid G. Threat Adjusting Security. 2018. Available from: https://eprint.iacr.org/2018/084.pdf

[55] Equivoe-T: Transposition Equivocation Cryptography. US Patent 10,608,814. 2020

[56] Samid G. SpaceFlip: Unbound Geometry Cryptography. 2019. Available from: https://eprint.iacr.org/2019/285.pdf

[57] Samid G. SpaceFlip: Unbound Geometry Cryptography. 2019. Available from: https://dblp.org/rec/journals/iacr/Samid19.html

[58] Spaceflip: Unbound Geometry Security. US Patent 10,790,977. 2020

[59] Samid G. A New Perspective of Geometry and Space as an Evolutionary Organizer of Data. 2017. Available from: http://www.dgsciences.com/Geometry_H7n18.pdf

[60] Samid G. Family Key Cryptography: Interchangeable Symmetric Keys; A Different Cryptographic Paradigm. 2021. Available from: https://eprint.iacr.org/2021/458

[61] SpaceFlip Plus: Ordinal Cryptography. US Patent 11,159,317. 2021

[62] Samid G. At-will intractability up to plaintext equivocation achieved via a cryptographic key made as small, or as large as desired—Without computational penalty. In: International Workshop on Cryptology and Network Security; September 26–28, 2002; San Francisco, California, USA

[63] BitMap Lattice: A Cyber Tool Comprised of Geometric Construction. US Patent 10,911,215. 2021

[64] Samid G. Denial Cryptography Based on Graph Theory. 2004. US Patent 6,823,068

[65] Samid G, Popov S. BitFlip: A Randomness Rich Cipher. 2017. Available from: https://eprint.iacr.org/2017/366.pdf

[66] BitFlip Cyber Demonstration. 2021. Available from: http://wesecure.net/learn/BitFlipEncrypt.php

[67] Transmitter for Encoding Information with Randomly Flipped Bits and Transmitting That Information Through a Communication Channel. US Patent 10,728,028. 2020

[68] Advanced BitFlip: Threat Adjusted, Quantum Ready, Battery Friendly, Application Rich Cipher. US Patent 10,541,808. 2020

[69] Samid G. A Unary Cipher with Advantages over the Vernam Cipher. 2020. Available from: https://eprint.iacr.org/2020/389

[70] Unary Cryptography Demonstration Site. 2020. Available from: https://UnaryCryptography.com

[71] Split Security Solutions. US Patent Application 17/510,324. 2021

[72] Randomized Bilateral Trust (RABIT): Trust Building Connectivity

for Cyber Space (FigLeaf). U.S. Patent 10,798,065. 2020

[73] Efficient Proof of Knowledge of Arbitrarily Large Data Which Remains Undisclosed. US Patent 10,594,480. 2020

[74] Method for Inhibiting Mass Credentials Theft. US Patent 10,395,053. 2019

[75] Proving Material Identity with Quantum Randomness – Financial and General Applications. US Patent 10,754,326. 2020

[76] Samid G. BitMint hard wallet: Digital payment without network communication: No internet, yet sustained payment regimen between randomness-verifiable hard wallets. In: IOT, Electronics and Mechatronics Conference (IEMTRONICS). New York City, NY: IEEE International; 2020

[77] Rock of Randomness. US Patent 10,467,522. 2019

[78] Samid G, Wnek G. The rock of randomness: A physical oracle for securing data off the digital grid. Material Research Society Bulletin. MRS Communications (Berlin: Springer). Mar 2019;**9**(1):67-76

[79] Marton K, Suciu A, Ignat I. Randomness in digital cryptography: A survey. Romanian Journal of Information Science. 2010. Available from: https://www.academia.edu/download/46676431/Randomness_in_Digital_Cryptography_A_Sur20160621-25262-h5ar54.pdf

[80] Quantum Random Number Generation. 2023. Available from: https://www.idquantique.com/random-number-generation/overview/

[81] Samid G. Randomness as absence of symmetry. In: The 17th International Conference on Information & Knowledge Engineering (IKE– 18): July 30-August 2, 2018; Las Vegas, USA. Available from: http://bitmint.com/SymRand_Vegas_H8518R.pdf

Section 2

Biometrics

Chapter 5

Introductory Chapter: The Fundamentals of Biometrics – A Comprehensive Exploration

Carlos M. Travieso-González

1. Introduction

1.1 Unveiling the future for biometrics

In the age of rapid technological advancement, the quest for secure and convenient identity authentication methods has never been more pressing. Traditional forms of authentication, such as passwords and PINs, are increasingly vulnerable to breaches, leaving individuals and organizations alike vulnerable to identity theft, fraud, and unauthorized access. In response to these challenges, biometrics has emerged as a promising solution, offering a unique fusion of security, convenience, and reliability [1]. From fingerprint scanners and facial recognition systems to iris scanners and voice authentication technologies, biometric systems leverage the distinct physiological and behavioral characteristics of individuals to verify their identities with unparalleled accuracy.

The term "biometrics" derives from the Greek words "bio" (life) and "metrics" (measurement), reflecting its focus on the measurement and analysis of biological traits. Unlike traditional forms of authentication, which rely on knowledge (e.g., passwords) or possession (e.g., ID cards), biometrics authenticates individuals based on who they are, leveraging unique physical or behavioral attributes that are inherently difficult to forge or replicate [2]. This paradigm shift has profound implications for a wide range of applications, from securing access to personal devices and financial accounts to enhancing border security and safeguarding critical infrastructure.

1.2 Biometrics modalities and its applications

At the heart of biometric technology lies the recognition and analysis of biometric identifiers, which can be broadly classified into two categories: physiological and behavioral [3]. Physiological biometrics are based on anatomical or physiological characteristics of individuals, such as fingerprints, iris patterns, facial features, and DNA profiles. These traits are inherently unique to each individual and remain relatively stable over time, making them ideal for reliable identity authentication. In contrast, behavioral biometrics are based on patterns of behavior or actions exhibited by individuals, such as typing rhythms, gait patterns, facial information [4–6], handwriting [7, 8], and voice characteristics. While behavioral biometrics may exhibit greater variability than physiological traits, they offer additional layers of security and can be less intrusive in certain applications.

The roots of biometrics can be traced back to ancient civilizations, where methods such as fingerprints and seals were used to authenticate individuals and documents. However, it was not until the advent of modern computing and digital imaging technologies that biometrics began to realize its full potential. The development of automated fingerprint identification systems (AFIS) in the 1970s marked a significant milestone, paving the way for the widespread adoption of biometric authentication in law enforcement and forensic applications. Since then, rapid advancements in sensor technology, machine learning algorithms, and data processing capabilities have fueled the proliferation of biometric systems across diverse domains.

One of the key strengths of biometrics lies in its ability to provide strong authentication while enhancing user convenience and experience [9]. Unlike passwords or PINs, which can be forgotten, stolen, or shared, biometric traits are inherently tied to the individual and cannot be easily compromised. This inherent convenience has fueled the integration of biometric authentication into a wide range of consumer devices, from smartphones and tablets to laptops and wearables. Today, millions of users around the world rely on biometrics to unlock their devices, authorize transactions, and access digital services with a simple touch, glance, or voice command.

Moreover, biometrics offers a compelling solution to the growing challenge of identity fraud and cybercrime. With traditional authentication methods proving increasingly vulnerable to sophisticated attacks, organizations are turning to biometric technologies to bolster their security posture and protect sensitive data and assets. Biometric authentication not only provides a higher level of assurance than traditional methods but also offers greater scalability and usability, enabling organizations to strike a balance between security and user experience.

However, the widespread adoption of biometrics also raises important questions and concerns regarding privacy, security, and ethical implications. As biometric data becomes increasingly pervasive and interconnected, issues such as data protection, consent, and surveillance have come to the forefront of public discourse. The collection, storage, and processing of biometric data raise unique challenges and risks, requiring careful consideration of legal and regulatory frameworks to safeguard individual rights and mitigate potential abuses.

In this comprehensive exploration of biometrics, we embark on a journey to unravel the intricacies of this transformative technology and its implications for society. Drawing upon insights from diverse disciplines, including computer science, engineering, psychology, law, and ethics, we delve into the evolution, applications, and implications of biometrics in the digital age. Through a synthesis of theoretical perspectives, empirical research, and real-world case studies, we seek to deepen our understanding of the promises and perils of biometric authentication and its role in shaping the future of identity management and security [10].

As we navigate the complex landscape of biometrics, we are reminded of the words of science fiction author Arthur C. Clarke, who famously remarked, "Any sufficiently advanced technology is indistinguishable from magic." Indeed, in the realm of biometrics, the boundaries between science and science fiction blur as we unlock the potential of human biology to transform the way we authenticate, interact, and connect in an increasingly digital world. Join us on this captivating journey into the realm of biometrics, where the future meets the present, and the possibilities are limited only by our imagination.

Author details

Carlos M. Travieso-González
Department of Signals and Communication, Institute for Technological Development and Innovation in Communications, University of Las Palmas, Gran Canaria, Spain

*Address all correspondence to: carlos.travieso@ulpgc.es

References

[1] Jain AK, Ross A, Nandakumar K. Introduction to Biometrics. Springer; 2016

[2] Wayman JL, Jain AK. Biometric Systems: Technology, Design, and Performance Evaluation. Springer Science & Business Media; 2005

[3] Maltoni D, Maio D, Jain AK, Prabhakar S. Handbook of Fingerprint Recognition. Springer Science & Business Media; 2009

[4] Li SZ, Jain AK. Handbook of Face Recognition. Springer Science & Business Media; 2005

[5] Almeida V, Dutta MK, Travieso CM, Singh A, Alonso JB. Automatic age detection based on facial images. In: 2016 2nd International Conference on Communication Control and Intelligent Systems (CCIS). Mathura, India; 2016. pp. 110-114. DOI: 10.1109/CCIntelS.2016.7878211

[6] Briceño JC, Travieso CM, Alonso JB, Ferrer MA. Robust identification of persons by lips contour using shape transformation. In: 2010 IEEE 14th International Conference on Intelligent Engineering Systems. Las Palmas, Spain; 2010. pp. 203-207. DOI: 10.1109/INES.2010.5483848

[7] Kutzner T, Travieso CM, Bönninger I, Alonso JB, Vásquez JL. Writer identification on mobile device based on handwritten. In: 2013 47th International Carnahan Conference on Security Technology (ICCST). Medellin, Colombia; 2013. pp. 1-5. DOI: 10.1109/CCST.2013.6922063

[8] Vargas JF, Travieso CM, Alonso JB, Ferrer MA. Off-line signature verification based on gray level information using wavelet transform and texture features. In: 2010 12th International Conference on Frontiers in Handwriting Recognition. Kolkata, India; 2010. pp. 587-592. DOI: 10.1109/ICFHR.2010.96

[9] Vásquez JL, Ravelo-García AG, Alonso JB, et al. Writer identification approach by holistic graphometric features using off-line handwritten words. Neural Computing and Applications. 2020;**32**:15733-15746. DOI: 10.1007/s00521-018-3461-x

[10] Ratha NK, Connell JH, Bolle RM. Enhancing security and privacy in biometrics-based authentication systems. IBM Systems Journal. 2001;**40**(3):614-634

Chapter 6

Revocable Crypto-Biometric Key Regeneration Using Face Biometrics, Fuzzy Commitment, and a Cohort Bit Selection Process

Mohamed Amine Hmani, Dijana Petrovska-Delacretaz and Bernadette Dorizzi

Abstract

Security is a main concern nowadays. Cryptography and biometrics are the main pillars of security. Using biometrics to obtain cryptographic keys offers distinct advantages over traditional methods. Classical systems rely on passwords or tokens assigned by administrators, which can be stolen or shared, making them insufficient for identity verification. In contrast, biometric-based keys provide a better solution for proving a user's identity. This chapter proposes an approach to regenerate crypto-biometric keys with high entropy, ensuring high security using facial biometrics. The keys are regenerated using a fuzzy commitment scheme, utilizing BCH error-correcting codes, and have a high entropy of 528 bits. To achieve this, we use an intra-inter variance strategy for the process of bit selection from our facial deep binary embeddings. The system is evaluated on the MOBIO dataset and gives 0% FAR and less than 1% FRR. The proposed crypto-biometric keys are resistant to quantum computing algorithms, provide non-repudiation, and are revocable and convenient with low false rejection rates.

Keywords: face biometrics, binarization, crypto-biometrics, key regeneration, key binding, fuzzy commitment, biometric cryptosystems, revocable crypto-biometric keys

1. Introduction

Cryptography has assumed an increasingly vital role in our society, driven by the digitization of all aspects of life and the escalating need to safeguard data and transactions across various domains such as telecommunications, healthcare, financial dealings, and even the protection of personal privacy, including cryptocurrencies. Cryptography principally relies on cryptographic keys to either encrypt data or create digital signatures, ensuring both data authenticity and integrity.

Nonetheless, cryptography faces certain challenges. One of its looming threats emanates from the advent of quantum computers. Quantum computer prototypes are becoming progressively more efficient and advanced, with the field of quantum computing progressing rapidly [1]. Projections suggest that within the next two decades, quantum computers could render current cryptographic methodologies obsolete. In this context, recommendations for post-quantum cryptographic algorithms, encompassing symmetric encryption, symmetric authentication, public-key encryption, and public-key signatures founded on long-term security, have been outlined in Ref. [2].

Beyond the quantum threat, cryptographic keys pose their set of inherent drawbacks. The sheer length of these keys precludes memorization, necessitating their storage, which introduces the additional risk of unauthorized key duplication or, worse yet, key loss resulting in the forfeiture of encrypted data. Traditional cryptographic keys also fail to guarantee non-repudiation, that is, the assurance that only the rightful owner uses the key, thereby enabling digital identity sharing and potential liability denial.

Biometrics emerges as a potential solution to this predicament. By employing biometrics, the issue of non-repudiation can be resolved. However, this approach gives rise to a new predicament: the irrevocability of biometric keys. It is impossible to arbitrarily alter an individual's biometric data, leading to the perpetual generation of the same biometric template for a given user.

Hence, our efforts focus on devising methods to create crypto-biometric keys derived from users' facial biometric data. We chose to focus on facial biometrics, given their widespread accessibility, as nearly every device nowadays is equipped with a camera. Employing a crypto-biometric key guarantees non-repudiation and eliminates the need for users to memorize the key, as it is reconstructed at the time of system usage.

The goal of this work is to obtain post-quantum crypto-biometric symmetric keys. This goal has multiple requirements:

1. to be resistant to quantum computing algorithms;

2. non-repudiation;

3. to be cancellable;

4. convenience, having low false rejection rate (FRR) at the required security level.

The rest of this chapter is structured as follows. In Section 2, we present an overview of related works in crypto-biometric key regeneration. In Section 3, we introduce the key regeneration scheme used in the proposed system. In Section 4, we provide the performance of the system. Before concluding in Section 6, we present the security analysis of the system in Section 5.

2. Related works

The concept of crypto-biometric keys appeared toward the end of the twentieth century. The first works had the disadvantage of low entropy and a non-negligible reconstruction error. Firstly, they demanded highly consistent biometric samples from the same user, which proved challenging due to environmental and physiological variations inherent to the nature of biometric data. Secondly, only one key was associated with a user's biometric sample, rendering the sample useless if the key was compromised, especially when periodic key updates were necessary.

To overcome these challenges, subsequent methods were introduced that did not require the storage of biometric templates or keys in databases using techniques such as key generation and key regeneration. Key generation is the process of deriving a key directly from biometrics. The keys can vary from a session to another, resulting in high FRR. As for key regeneration (also known as key binding [3]), it is the idea that a randomly generated key is combined with the biometric data using cryptographic techniques and that the key is later retrieved from the combined data at the time of verification. Sometimes, key generation and regeneration are used interchangeably, as the standards of crypto-biometric systems are not fully established.

In this section, we present related works in crypto-biometric key regeneration, focusing mainly on schemes using fuzzy commitment.

In 2007, Chen et al. introduced a method for regenerating cryptographic keys from facial biometrics in Ref. [4], utilizing entropy-based feature extraction and Reed–Solomon error-correcting codes. This technique proved capable of reliably producing keys suitable for 128-bit AES encryption when evaluated using 3D face data.

In 2010, Kanade et al. [5] proposed an iris biometric system with a key regeneration scheme based on fuzzy commitment use. Initially, a key is randomly generated and encoded into a pseudocode using error-correcting codes (ECCs). Subsequently, a cancellable transformation is applied to the user's reference biometric data, and this transformed data is XORed with the pseudocode to produce a locked code.

In 2018, Panchal et al. [6] introduced an innovative method for creating a threshold-free cryptographic biometric key, referred to as the crypto-biometric key. They achieved this by combining all bits from three different sets of straight line attributes obtained from three types of minutiae. Additionally, they employed Reed-Solomon encoding to generate a code word, which served the purpose of encryption and facilitated the retrieval of the user's crypto-biometric key. Notably, the authors asserted that their approach eliminated the need to store either the user's biometric template or the generated key. However, in Ref. [7], it is stated that the three feature sets may be the same set and their approach may have the risk of disclosing the fingerprint features. An attacker can apply the user's ciphertext to separate the code word from the user's final ciphertext and then obtain the features from the code word using Reed-Solomon decoding, and finally, he may successfully reconstruct the user's crypto-biometric key from the line attributes if they can obtain the parameters for substitution or expansion.

Anees et al. [8] have developed a unified framework for regenerating cryptographic keys of size 256 bits based on facial features, in 2018. There are three different modules in their proposed framework: learning the facial features, the quantization of the facial features, and key generation. In their application, the features of a subject are extracted in a controlled illumination with minimal variations in pose and expression at the time of enrolment and key regeneration.

In 2019, Panchal et al. [9] introduced a biometric code word generation technique based on the convolution coding principle using fingerprints. This method is capable of producing a distinct code word, and the hash value derived from this code word, in combination with specific private parameters, can serve as a cryptographic 1274-bit key.

In 2021, Wang et al. [7] proposed a new crypto-biometric key regeneration approach based on the generated interval scheme with a two-layer error-correcting technique for fingerprints. The key size in the proposed approach is adjustable between 120 and 168 bits long for the extracted minutia numbers with low computational cost.

Chang et al. [10] introduced a multi-biometric fusion system named BIOFUSE. They leverage the fuzzy commitment and fuzzy vault schemes to regenerate crypto-biometric keys of length 256 bits, combining face, iris, and fingerprint features.

Key regeneration can also function to verify the identity of the user; if the key is reconstructed correctly, the user is verified. Mai et al. [11] proposed a randomized convolutional neural network (CNN) to regenerate protected face templates given a facial image and a user-specific key. Their system was focused on biometric verification and provides 0.19% FRR @0.1% FAR on the FRGC database.

In 2022, Elrefaei et al. [12] proposed a fuzzy commitment scheme using gait-based biometrics to secure the feature templates. They leverage Bose–Chaudhuri–Hocquenghem codes (BCH) to encode the key in the enrolment phase and decode in the verification phase. They obtain 0% FAR and FRR on CMU MoBo and CASIA-A databases with key lengths of 50 bits.

Recently, in [13], Lin and Chen have proposed an error-correction-based iris recognition scheme that provides secure template storage and a high level of accuracy for personal authentication with a security of 110 bits, which means that the intruder requires 2^{110} attempts to access their system.

The previously mentioned works either focus on other biometrics modalities than face or have low crypto-biometric key sizes. Furthermore, in the works focusing on face biometrics, the testing databases had ideal conditions (illumination, pose), which reduces the convenience of the system and forces the user to comply with the system requirements. Moreover, the fuzzy commitment scheme was criticized for being vulnerable to correlation attacks and providing insufficient template security [14, 15]. Keller et al. [15], in particular, focused on the security of fuzzy commitment schemes based on biometric templates produced by deep learning.

In this chapter, we present a key regeneration approach that aims to regenerate long keys from facial data. Our focus is to have keys longer than 400 bits from facial biometric and working in noncontrolled environments. We employ a modified fuzzy commitment scheme that prevents template correlation and reconstruction. The following section presents the key regeneration scheme used in our system.

3. Crypto-biometric key regeneration with fuzzy commitment

In this section, we present our proposed key regeneration scheme. The system comprises two steps: (i) Extracting binary template from face biometrics and (ii) regenerating the crypto-biometric key using a fuzzy commitment scheme combining the face binary template, which is protected using shuffling protection scheme, and error-correcting codes. The first step is described in our paper [16] and summarized in Section 3.2. The challenge is to obtain long binary representations with high entropy and high biometric performance. These long binary representations are used as a basis for the second step. In the second step, the challenge is to regenerate the crypto-biometric keys without degrading the entropy and with low FRR.

The system proposed in this section is based on the fuzzy commitment scheme presented in **Figures 1** and **2**. The fuzzy commitment scheme was first introduced in 1999 by Juels and Wattenberg [17]. A random key is encoded using error-correcting codes (ECCs) and is then XORed with the biometric data. The XORed data is cryptographically secure because neither the key nor the biometric data can be obtained from it without providing one of the two. The random key is retrieved at the time of key regeneration by providing fresh biometric data. This system requires ordered biometric data in binary form. In this scheme, the variability in the biometric data from one acquisition to another is treated as noise. This noise causes errors in the data, which are corrected using ECC.

The system comprises two phases. The first phase, shown in **Figure 1**, is the enrolment phase where the user generates a symmetric key and links it to his/her identity. The second phase, shown in **Figure 2**, is the verification phase where the user regenerates his/her key from his biometric data, stored helper data, and a secret second factor used to ensure the revocability of the scheme.

The revocability of the fuzzy commitment scheme is assured using the same shuffling scheme described in section 3.4.

In the **enrolment phase**, the user provides an image containing his/her face *I* and a secret second factor *S*. The second factor can be a shuffling key stored on a secure token

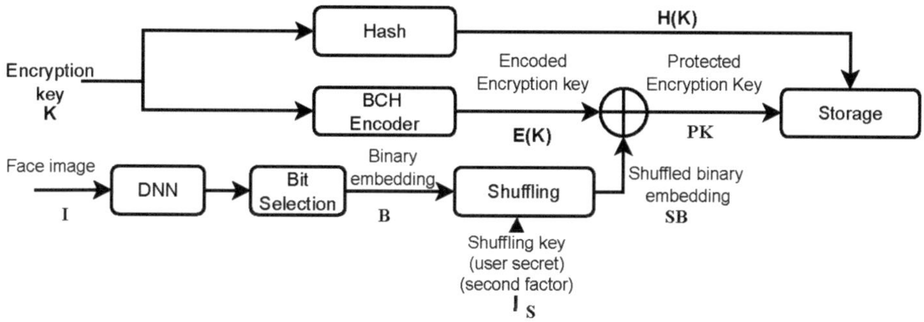

Figure 1.
Enrolment phase of the fuzzy commitment scheme used in the key regeneration.

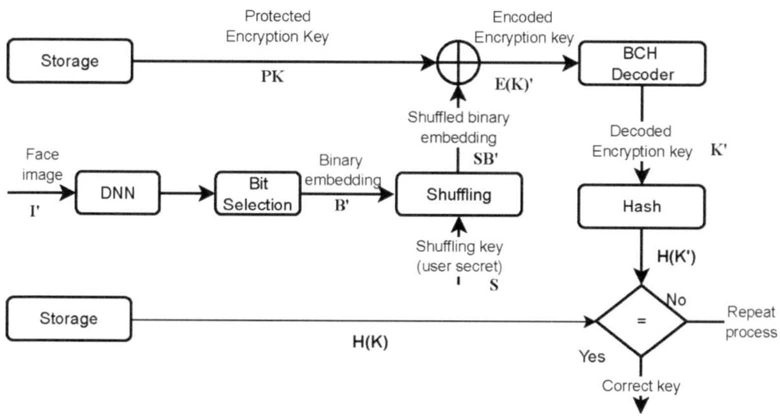

Figure 2.
Regeneration phase of the fuzzy commitment scheme.

or a password that is used to derive the shuffling key. The image I is processed using the binarization method described in the section 3.2 to provide a binary embedding B.

The binary embedding is then shuffled using the shuffling key S provided by the user to achieve the revocability requirement. The shuffled binary embedding is denoted by SB. The SB is used to protect the encryption key K generated by the system at the beginning of the process. The encryption key K is encoded using the error-correcting code described in the following section, creating an encoded encryption key $E(K)$. The encryption key K is also hashed using a hash function to provide a hash-string that will be used to verify the successful regeneration of the encryption key K in the regeneration phase.

The shuffled binary embedding SB and the encoded key $E(K)$ are XORed to provide the protected encryption key PK. The security of the scheme is mainly provided by the SB. In fact, the entropy of the system is the minimum between the entropy of SB and K.

In the enrolment phase, the system stores the hash of the encryption key $H(K)$ and the protected encryption key PK. The two pieces of data do not need to be protected and can be stored in a non-encrypted database. On the other hand, the second factor, S, must be protected. Should the shuffling key be compromised, the user's enrolment must be revoked and a new enrolment using a different shuffling key S needs to be generated.

In the **verification phase**, the user provides a new facial image sample I'. This image is processed following the same method as in the enrolment phase to provide a binary embedding B'. The user also provides the same second factor, in the form of either a password or a shuffling key S, used in the enrolment phase. The binary embedding B' is shuffled using this second factor, resulting in a shuffled binary embedding SB'. This shuffled binary embedding SB' will have some differences from SB due to the variability in the facial image sample provided at the beginning of the verification step. SB' is then XORed with the protected encryption key PK recovered from storage. The result of the XOR operation is $E(K')$. $E(K')$ is then decoded using the same error correcting code used in the enrolment phase. The decoded encryption key is denoted by K'. To check if the regeneration of the key is successful or not, we compare the hash of the decoded key $H(K')$ with the hash stored in the enrolment phase $H(K)$. If $H(K)$ is equal to $H(K')$, the regeneration is successful and the system provides the user with the encryption key K' that is identical to the key K. If $H(K)$ and $H(K')$ are different, the regeneration fails, and the user is asked to provide a new facial image.

The description of the fuzzy commitment scheme provided above explains only the generic data flow in the system. In our system, the binary embedding extractor (DNN) provides a fixed output of 4096 bits. To adapt the length of the binary embedding to the need of the security requirements of the system, we added a process of bit selection. In fact, according to the parameters of the error-correcting code and the length of the encryption key K, the length of the encoded encryption key $E(K)$ will be different. Therefore, the length of SB will vary as it hides the encoded encryption key $E(K)$. As such, the length of the binary embedding B and shuffling key S has to vary. The details about the shuffling scheme, the error-correcting code used, as well as the process of bit selection are provided in the following sections.

3.1 Error-correcting code

The difference between E(k) and E(k') is due to the variability in the biometric samples. We consider the difference as channel noise and use ECC to correct this noise. The error-correcting code that we used for the fuzzy commitment scheme is the Bose, Chaudhuri, and Hocquenghem (BCH) code. We chose the BCH code because it is a robust code capable of correcting random errors. BCH codes are a class of cyclic error-correcting codes that are based on algebraic concepts. They are widely used in communication systems and storage devices to detect and correct errors.

The basic idea behind BCH codes is to add redundant bits to the data being transmitted or stored in such a way that the receiver or reader can use these bits to detect and correct errors. These redundant bits are called check bits. The number of check bits added to the data is determined by the length of the BCH code and the desired error correction capability.

The BCH code is constructed using a generating polynomial, which is a polynomial equation with coefficients that are used to generate the check bits. The generating polynomial is chosen such that it has a certain number of roots, which are values that make the polynomial equation equal to zero. These roots are used to generate the check bits, and the number of roots determines the error correction capability of the BCH code. Each coefficient in the polynomial represents a bit of data, and the polynomial as a whole represents the entire dataset. The polynomial is then used to generate a set of check bits, which are added to the dataset to form the coded message.

To encode data using a BCH code, the data is first divided into blocks of a certain size, and the check bits are generated for each block using the generating polynomial. The check bits are then appended to the data block to form the encoded data block. When the encoded data block is transmitted or stored, errors may occur due to noise or other factors.

To detect and correct errors in the received or retrieved data, the receiver, or reader, uses the generating polynomial to calculate the check bits for the received or retrieved data block. If the calculated check bits match the check bits in the received or retrieved data block, it is assumed that the data is error-free. If the calculated check bits do not match the check bits in the received or retrieved data block, it is assumed that errors have occurred and the receiver or reader uses the generating polynomial to determine the locations of the errors and correct them.

One of the advantages of BCH codes is their robustness against errors. These codes can correct a certain number of errors based on their designed parameters and are therefore able to tolerate a certain level of noise or interference during the transmission or storage of data. This makes them particularly useful for applications where the transmission or storage of data may be subject to noise or interference.

In summary, BCH codes are a type of error-correcting code that are used to detect and correct errors in digital data. They are based on algebraic concepts and use a generating polynomial to generate check bits that are appended to the data being transmitted or stored. The receiver or reader uses the generating polynomial to detect and correct errors in the received or retrieved data. BCH codes have a high error correction capability and low overhead, making them efficient for use in communication systems and storage devices.

The BCH code takes a block of size $\boldsymbol{k}$ and encodes it on a code word of length $\boldsymbol{n}$. The code has a correction capacity of $\boldsymbol{t}$, meaning in each code word, we can correct at most $\boldsymbol{t}$ errors. The parameters $\boldsymbol{n}$, $\boldsymbol{k}$, and $\boldsymbol{t}$ are defined by Eq. (1).

For any positive integers $m \geq 3$ and $t < 2^m - 1$, there exists a binary BCH code with the following parameters:

$$\begin{array}{lll} \text{code word length} & n = 2^m - 1 & \\ \text{number of partiy check bits} & n - k \leq m \times t & (1) \\ \text{minimum distance} & d_{min} \geq 2t + 1 & \end{array}$$

with the minimum distance being the minimum number of positions in which any two distinct code words differ.

The encryption key of length L is divided in N blocks of k bits. Each block is encoded into a new block of n bits using the BCH error correcting code. The total length of the encoded key is $(N \times n)$. The scheme can correct up to $T = (N \times t)$ errors, where t is the correction capacity of the code.

3.2 Face binary template extraction

Crypto-biometric schemes such as fuzzy commitment require binary sources [18]. In Ref. [16], we use a data-driven template binarization method using deep neural networks (DNNs), which does not degrade the performance of the baseline system. Furthermore, we seek to obtain long binary representations with high entropy to be used in crypto-biometric key regeneration schemes. The binarization method is detailed in Ref. [16]. Using a pre-trained convolutional neural network (CNN) and training the model on a cleaned version [19] of the MS-celeb-1 M database, we obtain binary representations of length 4096 bits and 3300 bits of entropy. The extracted representations have high entropy and are long enough to be used in crypto-biometric systems such as fuzzy commitment. Furthermore, the proposed approach is data-driven and constitutes a locality-preserving hashing that can be leveraged for data clustering and similarity searches. Further details about the implementation are provided in Ref. [16].

3.3 Bit selection process

According to the length of the encryption key, the block size used, and the code word size of the ECC, the output encoded key $E(K)$ will have varying sizes. For example, if the length of the encryption key K is 512 bits, we divide it into 16 blocks of 32 bits. Each block is encoded on a code word of 63 bits. The encoded encryption key will have 16 words, resulting in a total length of 1008 bits. If the length of the encryption key is 516 bits (a 512-bit key padded with 4 zeros), the key can be divided into 86 blocks of 6 bits. Each block will be encoded on a 31-bit code word. The final output will have a length of 2635 bits.

However, the binary embedding extractor provides binary representations of a fixed length of 4096 bits. As the histogram in **Figure 3** shows, not all the bits of the binary representation have high entropy. As such, we proceed to select the bits with the most entropy of the binary representations. Furthermore, we need to preserve the recognition performance of the binary representations when selecting the bits.

The bit selection is made by first reordering the bits of the binary representation following the inter-class variance, intra-class variance, or both. Then, we take the first N bits needed in the fuzzy commitment scheme. The computation of the inter-class variance and intra-class variance is done on a cohort database that has no overlap with the validation database.

We selected the FRGC database [20] as the cohort database because of two reasons. First, it was not acquired in the wild like LFW [21] or MS-celeb-1 m [22].

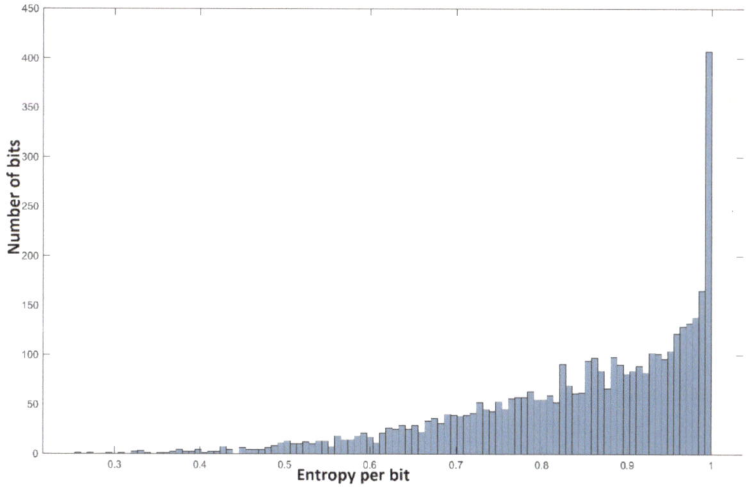

Figure 3.
Entropy per bit of the 4096-bit binary representations.

As such, there is a low risk of mislabeling or overlap with other databases. Secondly, it provides images of high quality in controlled conditions, which are useful for computing inter-class variance without introducing ambiguities due to the acquisition conditions. **Figure 4** shows examples of the images used to compute the inter-class variance. The images are frontal facing with good lighting and a uniform background. We also computed the intra-class variance using the FRGC dataset, but using the uncontrolled partition, as shown in **Figure 5**.

Figure 6 shows the process of reordering the bits using the inter-class variance computed using FRGC. We take a binary representation of a controlled sample from each subject in the database. Then, we compute the variance for each component (column) (considering that the bit distribution is a Bernoulli distribution and the bit probability is computed empirically). We then reorder variances using descending order. Following this process, we store the indices following the new order for future use.

In the key regeneration system, each time the binary representation is extracted, we use the stored indices to select the suitable number of bits for the parameters of the system. Using these indices, we can select the N bits needed for the scheme. The

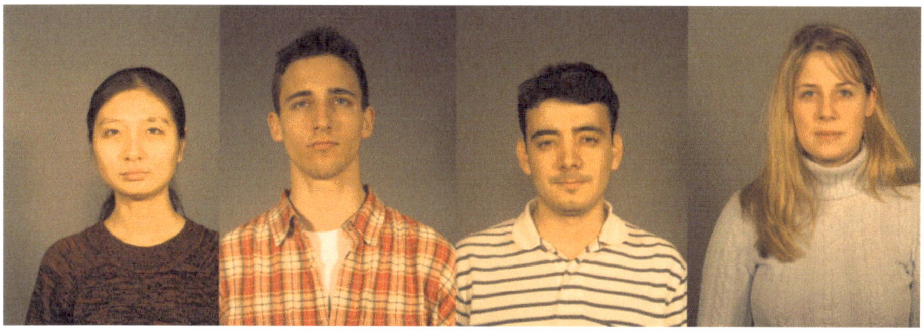

Figure 4.
Example of the images used to compute the inter-class variance. The images are taken from the controlled partition of the FRGC database [20].

Figure 5.
Example of the images used to compute the intra-class variance. The images are taken from the uncontrolled partition of the FRGC database [20].

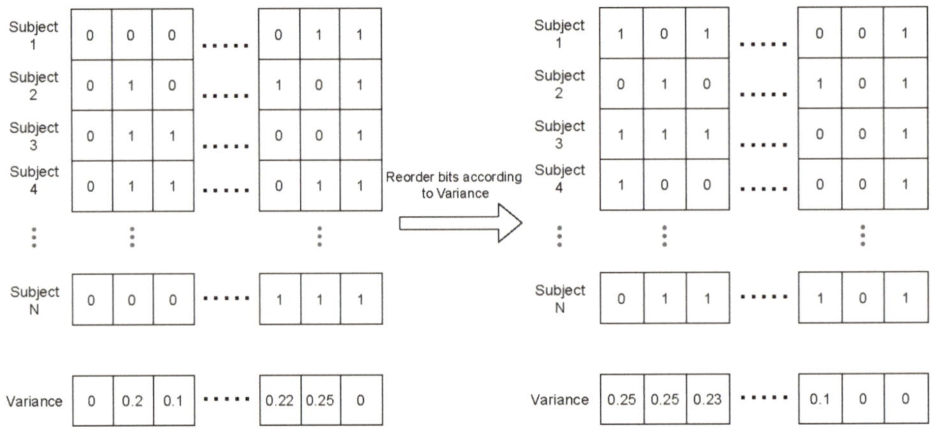

Figure 6.
Bit reordering of the binary representations created by the DNN according to the inter-class variance.

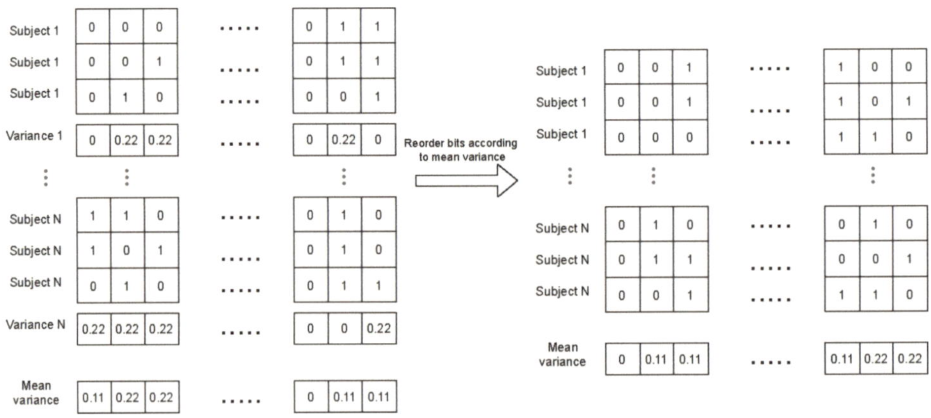

Figure 7.
Bit reordering of the binary representations created by the DNN according to the intra-class variance.

selected bits will be the bits with the highest entropy from the binary face representations. This is under the assumption that there is no session noise when the data was acquired. That is why we only computed the inter-class variance using only samples from the controlled partition.

As for the computation of the intra-class variance, we used the uncontrolled partition of the FRGC database. In this case, we take all the samples pertaining to each user. Then, we compute the variance for each user independently using all the samples. As a result, we obtain a variance vector for each user, as illustrated in **Figure 7**. The variance vectors are then averaged to provide the mean variance vector. In this mean variance vector, the bits with the highest variance are the bits that represent the session noise contained in the input samples. In this case, the variance vector is ordered using an ascending order. And same as the case of the inter variance, we store the indices of the new order.

As the goal of this type of bit selection is to remove the noisy bits, the need for using the FRGC database is further emphasized as we are confident that the data does not contain mislabeling.

Selecting the bits using the inter-class variance improves the security of the system at the cost of the convenience of the user. By taking the bits with the highest inter-class variance, we reduce the false acceptance rate, which increases the false rejection rate of the system. On the other hand, if we focus only on the bits with the lowest average intra-class variance, the user will have an easier time regenerating his/her key, but this will increase the risk of false acceptance. As such, we tried to combine both approaches by selecting the bits using both intra-class and inter-class variance.

Figure 8 illustrates the process of the bit selection process. We subtract the intra-class mean variance from the inter-class variance vector. If the bit has high inter-class variance and low intra-class variance, in the resulting vector, it will obtain a high weight. If the bit has low inter-class variance and high intra-class variance, it will get a low weight in the new vector. Finally, this new vector is ordered in descending order, and we take the first N bits needed in the fuzzy commitment scheme.

To validate the selection process, we used the accuracy on the LFW database as the benchmark. The tests carried out on the LFW database are divided into 3000 client-client tests and 3000 client-imposter tests. As such, the accuracy metric can give a

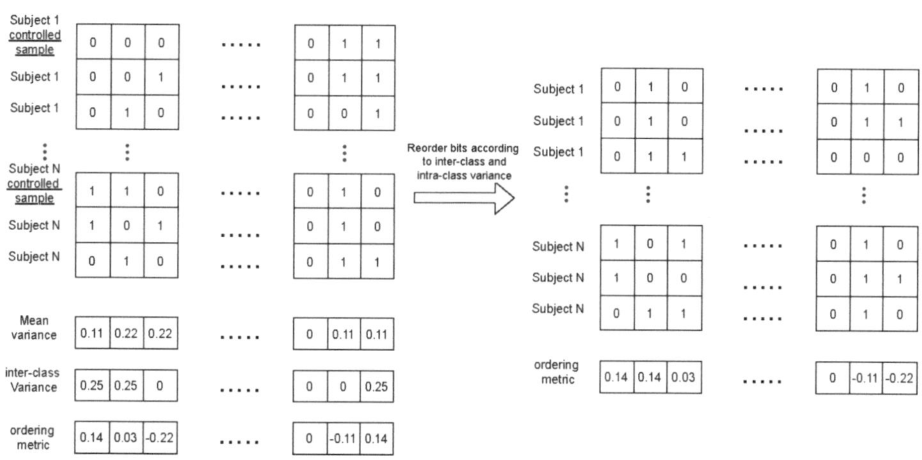

Figure 8.
Bit reordering of the binary representations created by the DNN using the inter-class variance and intra-class variance.

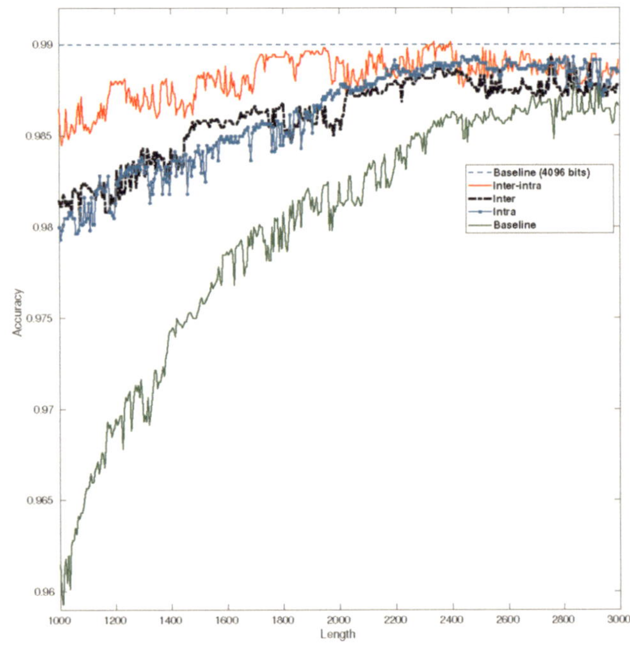

Figure 9.
Impact of the bit selection strategy on the accuracy of the binary representations on LFW [23]. The accuracy is represented as a function of the length of the representation.

sensible measure of the usability of the system. If the system has high accuracy, then it achieves both a low false acceptance rate and a false rejection rate.

Figure 9 shows the performance of the inter-class, intra-class, and inter-class + intra-class bit selection strategies. The figure shows the accuracy on LFW as a function of the length of the binary representation. All the curves are drawn using the same initial binary representation; we only changed the selection strategy. As the lowest length of the encoded message (encryption key) is above 1000 bits, the curves start at length 1000 bits, and each data point is computed using an increment of 10 bits.

The average size of the encoded keys in our experiments is between 2000 and 3000 bits. Thus, we chose the bit selection based on the third strategy, which is to use both the inter-class and intra-class variance to select the bits as this strategy has the best performance in this range as shown in **Figure 9**.

3.4 Biometric template protection scheme: shuffling

To protect template data, we adopt the shuffling scheme introduced by Kanade et al. in Ref. [24]. This scheme employs a binary shuffling key, which due to its length, is either securely stored on a dedicated token or derived through a password-based mechanism. The binary embedding, representing the template, is partitioned into blocks. This partition yields two distinctive segments: the first comprises blocks aligned with positions where the shuffling key exhibits a bit value of "1", while the second encompasses all remaining blocks. These two segments are subsequently concatenated, constituting the shuffled binary embedding. It is noteworthy that the original and shuffled templates exhibit a one-to-one correspondence; however, each block from the original vector has a different position within the shuffled embedding.

As a result, when identical shuffling keys are employed to shuffle two binary embeddings, the absolute block positions are altered consistently across both representations, leaving the Hamming distance unaffected. On the other hand, using distinct shuffling keys results in a randomized transformation of the initial binary representations, leading to an increased Hamming distance.

In the context of our application, namely, crypto-biometric key regeneration, we have opted for a block size of "1", rather than "7" employed in [24]. This choice provides two primary advantages. First, it generates a longer shuffling key, rendering it more resistant to brute-force attacks. Second, it expands the permutation space, providing a higher number of potential templates.

The shuffled binary embedding is the outcome of combining the biometric binary embedding with the shuffling key. Therefore, in the event of a security compromise, it can be revoked, and a new enrolment can be generated with the same biometrics using a different shuffling key. The management of shuffling keys involves two viable approaches: the generation and storage of keys within a secure element or their derivation from a password using mechanisms such as a password-based key derivation function (PBKDF). PBKDF, in essence, represents a mechanism for transforming a password into a symmetric key suitable for cryptographic operations, such as the advanced encryption standard (AES).

4. Results of the proposed key regeneration scheme on the MOBIO database

The proposed key regeneration scheme illustrated in **Figures 1** and **2** can be summarized as follows:

- **Enrolment phase**

    1. **Input:** The user presents a good-quality image of their face I and second factor S.
    2. The system generates an encryption key K.
    3. The system computes $E(K) = ENCODE_{BCH}(K)$ and $H(K) = secure\_hash\_function(K)$.
    4. The system extracts a binary representation from the facial image [16].
    5. The binary representation length is reduced to the desired length using the inter-intra variance bit selection process to obtain the binary embedding B.
    6. The binary embedding B is shuffled using the secret factor S to produce the shuffled binary embedding SB.
    7. The system computes the protected encryption key $PK = E(K) \oplus SB$.
    8. **Storage:** PK and $H(K)$ are stored in the system database, and all other data is discarded.

The second type of attack is to try to use protected encryption keys to recover the original encryption key. In this case, the attacker tries to XOR the protected encryption key with binary strings generated using the metadata of the system. We suppose that attackers know the statistical distribution of the binary face representations and how the shuffling keys are generated. The shuffling keys can be either randomly generated or stored on a special medium for the user to use (a smart card) or derived using a PBKDF from a password provided by the user. Using this information, the attacker can reduce the complexity of his attack by knowing the average number of activated bits in the shuffled binary embeddings *SB* used to protect the encoded encryption keys $E(K)$. In this attack, the goal of the adversary is mainly to generate a "pseudo" shuffled binary embedding to reverse the XOR operation applied to the encoded encryption key $E(K)$. The complexity of the attack is equivalent to the diversity of the cancellable system computed using the parameters of the fuzzy commitment scheme. The maximum number of shuffled binary embeddings *SB* is given using the number of possible permutations. Moreover, because the decision-making is based on a threshold comparison, we should not account for templates falling in the same neighborhood. We estimate the maximum number of templates using the hamming-packing bound.

$$Number\ Of\ SB = \frac{number\ of\ permutation}{volume\ of\ Hamming\ spheres} = \frac{L!}{L_0!L_1!\sum_{k=0}^{t}\binom{L}{k}} \tag{2}$$

where:
L: the length of the encoded encryption key/shuffled binary representations.
L_0: the average number of zeros in the shuffled binary representations.
L_1: the average number of ones in the shuffled binary representations.
t: the maximum of number of bits that can be corrected using the ECC used in the fuzzy commitment scheme, where $2t + 1$ is the minimum distance of the ECC.

For example, in the case where we use encryption keys of length 528 bits with a BCH(127,22,23) code with a correction capacity of 23 bits in a block of 127 bits of the encoded message, the encryption key is divided into 24 blocks of 22 bits. Each block is encoded onto a 127-bit block. Thus, the maximum number of bits that can be corrected is 23×24. However, this does not mean that we can correct 552 random errors in the encoded message. We can correct only 23 random bits in each 127-bit block of the encoded message. As such, the number of possible *SB* in this case is higher than the lower bound provided by Eq. (2). We obtain more than 2^{992} possible *SB* for this configuration, as shown in Eq. (3).

$$Number\ Of\ SB = \frac{3084!}{1542!1542!\sum_{k=0}^{23\times 24}\binom{3084}{k}} \approx 2^{992} \tag{3}$$

We show in **Table 4** the minimum number of possible *SB* for each configuration of the system. We study the number of *SB* as it is equivalent to the entropy of the shuffled binary representations used to protect the encryption keys. For the case of

Encryption key length	BCH code	Number of *SB* ($\log_2$)
528	(127,22,23)	992
516	(31,6,7)	610
512	(63,32,11)	333
510	(63,10,13)	852
430	(2047,430, 214)	1056
430	(4095, 495, 430)	1916
420	(127,15,27)	901

The number of SB is provided in $\log_2$ *format. BCH codes are presented in* (n, k, t) *format where* n *is the length of the encoded block,* k *is the length of the message block, and* t *is the number of bits that can be corrected in the encoded block.*

Table 4.
Number of possible SB for each system configuration.

BCH (63,32,11), the complexity of the brute-force attack is reduced from 512 to 333, which is still not brute-forceable with current technology but lower than the security requirement that we established in Section 1. As for the rest of the configurations presented in **Table 4**, except for BCH (63,32,11), it is computationally infeasible to attack the system using the information recovered from the protected encryption key. It is easier to try to brute-force the encryption key directly using its signature.

5.4 Brute-force attacks

In this section, we study the brute-force attacks that can be done on the system. The attacker can try to either directly brute-force the encryption key K or brute-force the system by presenting random faces and random second factors to recover K.

The first attack is computationally infeasible for all the system configurations presented in **Table 1**. The encryption key lengths are longer than 400 bits, which are not brute-forceable even using quantum algorithms such as the Grover algorithm [28].

As for brute-forcing the system using its inputs (face and shuffling key), this attack is more complex than brute-forcing the encryption key as the shuffling key is longer than the original encryption key K. Furthermore, if we consider the best case for the attacker when they have biometric samples of the target, the attack reverts to the stolen biometric scenario, which is not brute-forceable.

6. Conclusion and perspectives

The challenge of this work is to obtain high entropy keys to guarantee a high level of security. In fact, for traditional cryptographic keys, one can control the entropy during their creation. On the other hand, with crypto-biometric keys, the entropy contained in the biometric reference limits the entropy of the key.

In this chapter, we propose a key regeneration scheme based on fuzzy commitment, deep binary embeddings, and a bit selection process based on cohort inter-intra variance. The regenerated keys are post-quantum, guarantee non-repudiation, and are revocable and convenient to use.

As we focus on symmetric key regeneration, the keys need to have double the entropy of the keys used currently to present the same degree of security [2]. Crypto-biometric keys are limited by the usable information contained in the biometric sample that they are generated from. This is the reason we use deep learning to create binary representations that are longer and more discriminative. State-of-the-art key regeneration systems that use face biometrics suffer from high FRR and low entropy compared to other biometric modalities [7]. In our case, we succeeded in regenerating symmetric encryption keys longer than 400 bits with low FAR and low FRR using face biometrics. For example, for crypto-biometric keys with 512 bits, we get 0% FAR and 0.3% FRR using BCH (63,32,11) code.

As for the revocability of the keys, biometrics are unique for each user. They cannot be changed without special circumstances (plastics surgery, diseases, and so on). As such, if the regeneration scheme is not revocable, the user will be restricted to a single key across multiple applications. In addition, in the case the key is compromised, the user will not be able to create a new one. Thus, to ensure that the regeneration scheme is revocable, we used the shuffling scheme proposed by [24].

The non-repudiation requirement is satisfied by the intrinsic properties of biometric samples (under the assumption that an anti-spoofing system is used). However, we must ensure that the scheme used in the key regeneration has a low false acceptance rate (FAR). In our case, we obtain 0% FAR for all configurations of the system on the MOBIO database.

Moreover, the proposed key regeneration scheme allows for user convenience, meaning, at the required security level, the user is not rejected multiple times before having access to the system. The convenience of the system is shown through the FRR metric (less than 1% FRR).

Furthermore, our system does not suffer from the security concerns presented in Ref. [15] about the unlinkability, irreversibility, and revocability of fuzzy commitment schemes, as we use a second factor for the protection of the biometric template. We have shown in Ref. [16] that the shuffling scheme is fully unlinkable and irreversible, provided that the user's secret second factor is not compromised. The revocability is also provided by means of the shuffling key. We can re-enroll the user using a different shuffling key.

Finally, the system can be further improved by using newer face recognition systems with higher accuracy as the basis for the binarization auto-encoder. This should allow for more stable and longer representations and as such longer crypto-biometric keys.

Abbreviations

FRR	false rejection rate
FAR	false acceptance rate
PBKDF	password-based key derivation function
ECC	error-correcting code
DBR	direct binary representation
BRGC	binary reflected gray code
LSSC	linearly separable sub-code
BCH	Bose, Chaudhuri, and Hocquenghem
DNN	deep neural network
CNN	convolutional neural network

DOI: http://dx.doi.org/10.5772/intechopen.1003710

Author details

Mohamed Amine Hmani, Dijana Petrovska-Delacretaz* and Bernadette Dorizzi
Laboratoire SAMOVAR, Télécom SudParis, Institut Polytechnique de Paris, Palaiseau, France

*Address all correspondence to: dijana.petrovska@telecom-sudparis.eu

References

[1] Pooja S, SK. Quantum computing review: A decade of research. IEEE Transactions on Engineering Management. 2023:1-15. DOI: 10.1109/TEM.2023.3284689

[2] Augot D, Batina L, Bernstein DJ, Bos J, Buchmann J, Castryck W, et al. Initial recommendations of long-term secure post-quantum systems. In: Post-Quantum Cryptography for Long-Term Security. Eindhoven, Netherlands: PQCRYPTO; 2015. Available from: pqcrypto.eu.org/docs/initial-recommendations.pdf

[3] Jain AK, Ross A, Uludag U. Biometric template security: Challenges and solutions. In: 2005 13th European Signal Processing Conference in Antalya, Turkey. New York City, USA: IEEE; 2005. pp. 1-4

[4] Chen B, Chandran V. Biometric based cryptographic key generation from faces. In: 9th Biennial Conference of the Australian Pattern Recognition Society on Digital Image Computing Techniques and Applications (DICTA 2007) in Glenelg, SA, Australia. New York City, USA: IEEE; 2007. pp. 394-401. DOI: 10.1109/DICTA.2007.4426824

[5] Kanade S, Petrovska-Delacrétaz D, Dorizzi B. Generating and sharing biometrics based session keys for secure cryptographic applications. In: 2010 Fourth IEEE International Conference on Biometrics: Theory, Applications and Systems (BTAS) in Washington, DC, USA. New York City, USA: IEEE; 2010. pp. 1-7. DOI: 10.1109/BTAS.2010.5634545

[6] Panchal G, Samanta D. A novel approach to fingerprint biometric-based cryptographic key generation and its applications to storage security. Computers and Electrical Engineering. 2018;**69**:461-478. DOI: 10.1016/j.compeleceng.2018.01.028

[7] Wang P, You L, Hu G, Hu L, Jian Z, Xing C. Biometric key generation based on generated intervals and two-layer error correcting technique. Pattern Recognition. 2021;**111**:107733. Available from:. DOI: 10.1016/j.patcog.2020.107733

[8] Anees A, Chen YPP. Discriminative binary feature learning and quantization in biometric key generation. Pattern Recognition. 2018;**77**:289-305

[9] Panchal G, Samanta D, Barman S. Biometric-based cryptography for digital content protection without any key storage. Multimedia Tools and Applications. 2019;**78**:26979-27000

[10] Chang D, Garg S, Ghosh M, Hasan M. BIOFUSE: A framework for multi-biometric fusion on biocryptosystem level. Information Sciences. 2021;**546**:481-511. Available from: https://www.sciencedirect.com/science/article/pii/S0020025520308318

[11] Mai G, Cao K, Lan X, Yuen PC. SecureFace: Face template protection. IEEE Transactions on Information Forensics and Security. 2021;**16**:262-277

[12] Elrefaei LA, Al-Mohammadi AM. Machine vision gait-based biometric cryptosystem using a fuzzy commitment scheme. Journal of King Saud University - Computer and Information Sciences. 2022;**34**(2):204-217. Available from: https://www.sciencedirect.com/science/article/pii/S1319157819300916

[13] Lin KC, Chen YM. A high-security-level iris cryptosystem based on fuzzy commitment and soft reliability extraction. IEEE Transactions on Dependable and Secure Computing. 2023:1-15. DOI 10.1109/TDSC.2023.3289916

[14] Tams B. Decodability Attack against the Fuzzy Commitment Scheme with Public Feature Transforms. New York, USA: arXiv; 2014. DOI: 10.48550/arXiv.1406.1154

[15] Keller D, Osadchy M, Dunkelman O. Fuzzy Commitments Offer Insufficient Protection to Biometric Templates Produced by Deep Learning. CoRR. 2020;**abs/2012.13293**. arXiv. Available from: https://arxiv.org/abs/2012.13293. Eprint: 2012.13293

[16] Hmani MA, Petrovska-Delacrétaz D, Dorizzi B. Locality preserving binary face representations using auto-encoders. IET Biometrics. 2022;**11**(5): 445-458. DOI: 10.1049/bme2.12096

[17] Juels A, Wattenberg M. A fuzzy commitment scheme. In: Proceedings of the 6th ACM Conference on Computer and Communications Security. New York, USA: Association for Computing Machinery (ACM); 1999. pp. 28-36

[18] Lim M, Teoh ABJ, Kim J. Biometric feature-type transformation: Making templates compatible for secret protection. IEEE Signal Processing Magazine. 2015;**32**(5):77-87

[19] Hmani MA, Mtibaa A, Delacretaz DP. Joining forces of voice and facial biometrics: A case study in the scope of NIST SRE19. In: Voice Biometrics: Technology, Trust and Security. Chapter 9. London, England: IET Digital Library; 2021. pp. 187-217. Available from: https://digital-library.theiet.org/content/books/10.1049/pbse012e_ch9. DOI: 10.1049/PBSE012E_ch9

[20] Phillips PJ, Flynn PJ, Scruggs T, Bowyer KW, Chang J, Hoffman K, et al. Overview of the face recognition grand challenge. In: 2005 IEEE Computer Society Conference on Computer Vision and Pattern Recognition (CVPR'05). Vol. 1. London, England: IEEE; 2005. pp. 947-954

[21] Learned-Miller E, Huang GB, Roy Chowdhury A, Li H, Hua G. Labeled faces in the wild: A survey. In: Advances in Face Detection and Facial Image Analysis. New York, USA: Springer; 2016. pp. 189-248

[22] Guo Y, Zhang L, Hu Y, He X, Gao J. Ms-celeb-1m: A dataset and benchmark for large-scale face recognition. In: European Conference on Computer Vision. New York, USA: Springer; 2016. pp. 87-102

[23] Erik LM, Gary BH, Aruni R, Haoxiang L, Hua G. LFW: Results. Amherst, MA, USA: Computer Vision Research Laboratory, University of Massachusetts. 2019. Available from: http://vis-www.cs.umass.edu/lfw/results.html.

[24] Kanade SG. Enhancing information security and privacy by combining biometrics with cryptography [thesis]. Lyon, France: Institut National des Télécommunications; 2010. Available from: https://tel.archives-ouvertes.fr/tel-01057728

[25] Bourlai T. Face recognition in challenging environments: An experimental and reproducible research survey. In: Face Recognition across the Imaging Spectrum. New York City: Springer; 2016. pp. 269-270

[26] Samaria FS, Harter AC. Parameterisation of a stochastic model for human face identification. In: Proceedings of 1994 IEEE Workshop on Applications of Computer Vision. New York, USA: IEEE; 1994. pp. 138-142

[27] Messer K, Matas J, Kittler J, Luettin J, Maitre G, et al. XM2VTSDB: The extended M2VTS database. In: Second International Conference on Audio and Video-Based Biometric Person Authentication in Washington, DC, USA. Vol. 964. Lausanne, Switzerland: Citeseer; 1999. pp. 965-966

[28] Grover LK. A fast quantum mechanical algorithm for database search. In: Proceedings of the Twenty-Eighth Annual ACM Symposium on Theory of Computing. New York, USA: Association for Computing Machinery (ACM); 1996. pp. 212-219

Chapter 7

Advanced Lightweight Encryption Key Management Algorithms for IoT Networks

Menachem Domb

Abstract

An Internet of Things (IoT) Network is a collection of sensors interconnected through a network that process and exchange data. IoT networks need sufficient resources to cope with the growing security challenges. In most cases, cryptography is implemented by symmetric and asymmetric encryption methods to cope with these security issues. Symmetric cryptography requires transmitting an encryption key to the receiver to decrypt the received encrypted messages. Consequently, secured key distribution techniques are the core for providing security and establishing a secured connection among objects. Encryption keys are frequently changed through key distribution mechanisms. Encrypted key exchange is a protocol that allows two parties who share the same key to communicate over an insecure network. This chapter outlines the challenges and core requirements for a robust key distribution mechanism, beginning with evaluating existing solutions and then detailing three innovative, efficient, and lightweight methods that balance the security level, network performance, and low processing overhead impact.

Keywords: key management/distribution, symmetric/asymmetric encryption, IoT networks, lightweight RSA, probability-based keys sharing

1. Introduction

IoT devices collect and distribute massive transactions and data in real time non-stop, which requires some means to secure this data, such as data encryption.

Figure 1 depicts the IoT's pivotal role in the complete picture of computing and communications. IoT networks comprise a wide range of interconnected devices that collect and analyze environmental data and act using actuators. IoT networks are utilized in various sectors, including smart energy grids, industrial control systems, healthcare, transportation, home appliances, and wearables [1]. The evolving IoT array adds numerous devices to the Internet with poor security resistance, risking the entire community of Internet users.

Symmetric Encryption methods are the best-balanced solution for such an enormous load, as they are reliable and have a minimal performance impact. However,

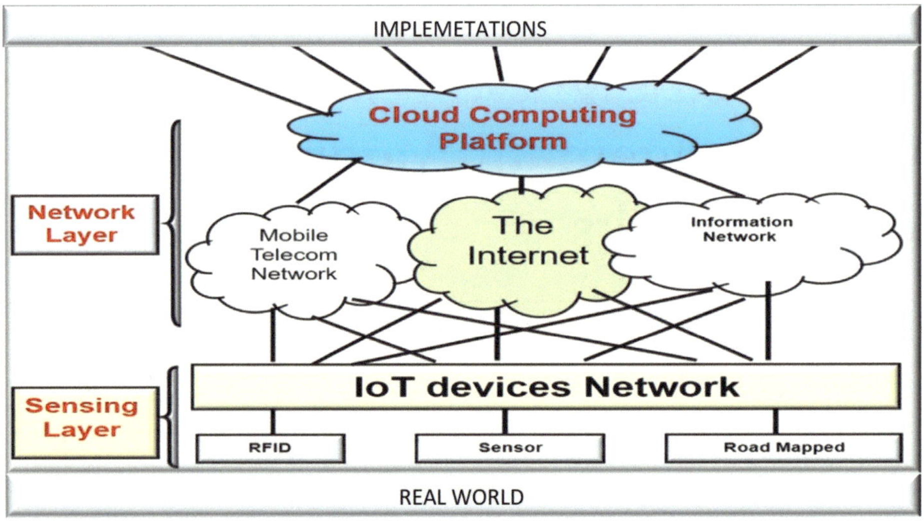

Figure 1.
IoT's role in the complete picture of computing and communications.

advanced processing capabilities reveal encryption keys instantly, forcing frequent key generation and spreading to have a unique encryption key per conversation, including generating, storing, distributing, and backing up the keys. IoT devices suffer from inherent weaknesses due to limited computing and communication resources, which prevent using standard key management systems designed for common networks. Dynamic key management schemes have already been proposed assuming homogeneous network architecture, while IoT networks are heterogeneous with no established standards. In this section, we suggest an ongoing key management process based on a probability analysis providing a key shared between any pair of IoT devices. The key size, randomness, sequence, and the number of alternate keys prevent attacks.

A comprehensive survey by Oraib et al. [2] outlines the most known key distribution methods suitable for IoT. They propose criteria to evaluate any key distribution schemes, which will be used later to compare various techniques in terms of performance and efficiency. The implementation should be scalable, resilient, and connective, while the efficiency concerns the node's communication, computing, and memory storage complexity. Some authors classified the key distribution based on the current proposals into trusted-server schemes and self-enforcing schemes. Hamid et al. [3] present four key distribution schemes, probabilistic, deterministic, hybrid, and group-based. Others classified the symmetric key distribution schemes in IoT into probabilistic, deterministic, and other categories. We propose to classify the secured key distribution into three classes, i.e., lightweight, robust encryption such as RSA, secure distribution by additional means such as internal CA, and key distribution means.

The chapter is organized as follows. Section 2 details the research related to secured key distribution and management. In sections 3, 4, and 5, we elaborate on the original and unique key distribution methods tailored to cope with sensor constraints, and in Section 6, we provide our conclusions.

2. Literature review

Naoui et al. [4] studied WSN key distribution protocols, mentioning centralized and decentralized methods, and concluded that a few proposals are comprehensive for different IoT applications. N-tier modeling of robust key management and cost-effective security paradigm with a 2-tier model to safeguard cloud data with effective authentication are addressed in [5, 6]. As the introduction explains, many solutions to the key distribution problem are classified into three categories. Below we provide a literature review for each of the methods.

2.1 Key distribution using lightweight encryption methods

A well-known key distribution method encrypts the distributed key using Asymmetric cryptography, such as RSA. However, Asymmetric encryption requires computation resources beyond a typical IoT device's resources. Therefore, a lightweight, fast, low computational cost algorithm is needed. Fadhil and Younis [7] discuss Elliptic Curve Diffie-Hellman (ECDH) key agreement algorithms for IoT. Goyal and Sahula [8] proposed a lightweight encryption algorithm for IoT devices using ECDH and AES encryption. Usman et al. [9] proposed an adaptive symmetric encryption algorithm (SIT) that merges Feistel and SPN with a 64-bit cipher. Nandini and Vanitha [10] analyzed several lightweight cryptography algorithms, such as HISEC, PRINCE, OLBCA, PRESENT, PRINT, TWINE, and KLEIN, and concluded that adding more S-boxes increases security. Some researchers implemented authentication by adding a new device as the authenticator. Ummer Iqbal et al. [11] proposed a lightweight ECC-based key exchange mechanism for fog federation. Their analysis indicates that it is safe from various attacks, with an overhead of 210.66 mJ and a communication overhead of 2144 bits, while conforming to the desired security specifications.

@@Eldefrawy et al. [12] presented a lightweight key distribution protocol for Industrial IoT that requires a single message exchange, handles node addition and cancelation, and fast rekeying. The scheme provides forward/backward privacy and avoids node capture and server takeoff attacks. Lian et al. [13] suited the traditional password-authenticated key exchange (PAKE) method to the IoT, with limited computing capability allowing two parties with a shared password to establish a session key. Their proposed protocol requires only three exponentiations per party while ensuring that the transmitted record and password file will not reveal the identity information. The Diameter protocol scheme provides a secure key agreement protocol that uses the ECDSA and the ECDH key agreement algorithm with less computational efforts suitable for IoT.

2.2 Key distribution assisted by external authentication means

Salman et al. [14] required a device to become a Certificate Authority (CA) server. Shivraj et al. [15] used cloud applications for constructing and distributing OTP. Aman et al. [16] proposed adding a separate device as a central authorization server and used simulations, BAN logic, and the Random Oracle model to assess its security strength simulations were performed for security verification. A detailed comparison of the proposed scheme with LKSE was conducted. Guo et al. [17] proposed a new AP-SGKD protocol using double chains and access polynomials. The new protocol is the AP-SGKD protocol that fulfills basic security properties with optimal storage requirements. Their simulation results showed that the new scheme could be applied to the Zigbee network since it performs well on security, storage, and communication. IoT nodes are connected to the Internet and communicate over a virtual network. No nodes with malicious intent are connected to the web to avoid cyberattacks. Moharana SR et al. [18] proposed a framework for the security over the virtual network for IoT nodes in a cloud system, including a lightweight cryptographic technique involving a key exchange protocol to establish secure end-to-end communication among the IoT nodes. This framework is a unique key exchange protocol between the CSP and the user group with the IoT nodes, which utilizes a balanced incomplete block design (BIBD) model.

Furthermore, it uses two communication channels, the Elliptic Curve Diffie Hellman (ECDH) protocol and the identity information for key exchange and sensor data communication. The ECDH generates the same shared secret key for the participants. Perfect Forward Secrecy (PFS) guarantees the safety of the hidden keys even if the private key is compromised. The secret key is derived using a hash function which is used later as the Key on Advanced Encryption Standard (AES) algorithm. The AES secures the sensor's data transmitted over the network.

2.3 Other key distribution methods

Orieb Abu Alghanam et al. [2] propose H2KD, a hierarchical architecture, and protocol for key distribution in IoT/WSN, which supports mobility, scalability, heterogeneity, and constrained nodes' limited capabilities. The performance was evaluated based on a quantitative measure of several metrics, memory storage, computation cost, scalability, the number of messages exchanged, and resilience required to establish a new session key for a mobile node. The results of their experiments show that the protocols are safe against attacks and reduce communication, computation, and storage costs for constraint nodes. The key agreement scheme uses the elliptic curve algorithm, and the symmetric encryption scheme uses AES and RC4. A quadratic-based wireless sensor key management scheme builds a shared key with a binary t-order symmetric polynomial, introduces a multivariate asymmetric quadratic polynomial, and utilizes the relationship between the quadratic eigenvalues and eigenvectors. It improves the anti-capture property, connectivity, scalability, communication overhead, and storage overhead. It is based on the Diameter protocol and introduces a new Z-Wave application layer protocol to provide end-to-end security. Othman et al. [19] propose a new AP-SGKD protocol using double chains and access polynomials. Their new protocol is the first AP-SGKD protocol that satisfies all basic security properties with an optimal storage requirement. In addition, they propose balancing the session key recovering time for less communication cost. Their simulation results show that their new scheme can be applied to the Zigbee network since it performs well on security, storage, and communication.

3. Probability-based keys sharing

This approach assumes a two-stage process. (i) Constructing a central pool of encryption keys and (ii) distributing sub-pools to the IoT devices. When two IoT devices want to establish communication, they identify a shared key, encrypt the message, and send it. These stages are executed within the IoT network devices.

The key-pool construction process is performed in a distributed and parallel mode. One IoT device is designated the Master, which requests the other devices to generate keys sent back and piled by the Master into one pool. The Master then randomly distributes keys to each IoT device such that any two devices have at least one shared key. **Figure 2** depicts the overall process stages.

We automated scalability and node mobility independent of this process in Python using Raspberry Pi-3 devices as IoT devices network. **Figure 3** illustrates the result of the key-pool set scattered into subsets where each subset has at least one shared key with another subset. Subsets are designated by the letter R and its key references. For example, nodes R3 and R4 share two keys, 8 and 5, and R4 and R5 share one reference key, 7. If there is no overlapping between two subnets, their communication is blocked. The implementation proves the feasibility of our proposed security protocol for IoT networks. The proposed scheme is symmetric or asymmetric and has network scalability and node mobility independent of the cryptography method. Assuming the keys are divided into subsets and distributed to all IoT network members. Each key has an assigned sequence number for security purposes, which will be used for inter-device communications. When a node is about to exchange messages with another node, it sends its list of key references. The receiving node intersects it with its subset, selects one of the intersected keys, and replies with the reference number chosen. The sender encrypts its message using the referenced key and sends it.

Eschenauer and Gligor [1] refer to a different setup, used a random graph, and employed probability-based key sharing. This setup does not contender with the unique IoT constraints. In [20], they used KMS and Asymmetric cryptography for IoT networks. Ciancalepore et al. [21] propose a Key Management Protocol for mobile and industrial IoT systems, with robust key negotiations, lightweight node authentication, fast rekeying, and efficient protection against replay attacks. It leverages ECC constructions, key exchange, and implicit certificates. Its advantage is that it allows suitable integration in a security protocol exchange such as 802.15.4. Roman et al. [22]

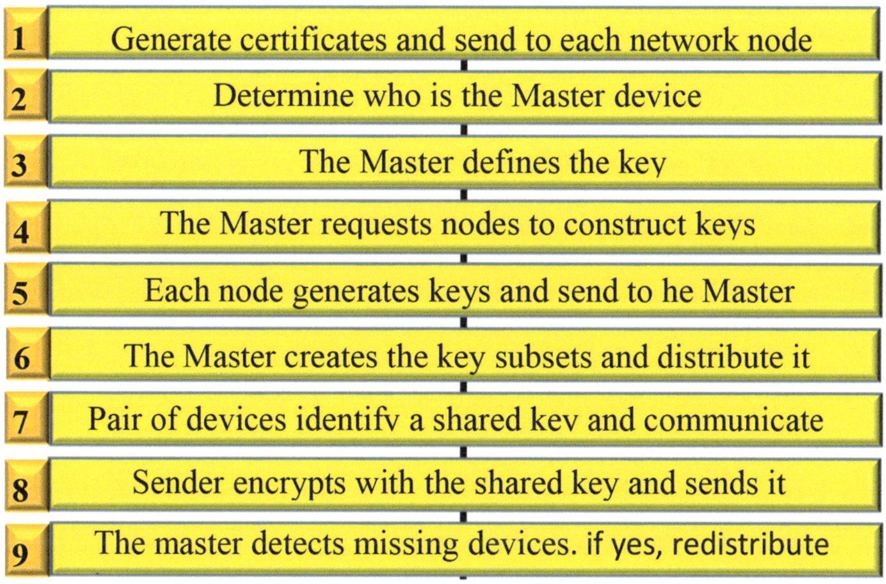

Figure 2.
Probability-based keys sharing stages.

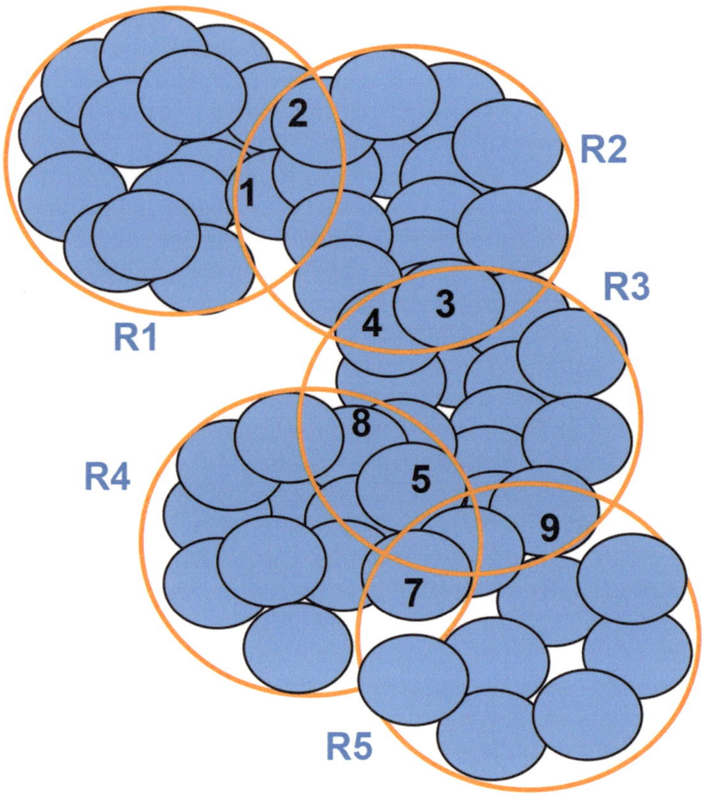

Figure 3.
The key pool after its division into subsets and before distribution to the devices.

propose key management mechanisms enabling two remote devices to negotiate specific security credentials while providing shared keys for sensors in the same network.

Wazid et al. [23] designed a new secure, lightweight three-factor remote user authentication scheme for IoT, using automated validation of Internet security protocols, offering offline sensing node registration and anonymity. Benslimane and BenAhmed [24] propose a lightweight key management protocol that allows the constrained node to transmit captured data to an internet host on a secure channel. Mahmood and Ghafoor [25] propose an Efficient Key Management (EKM) scheme for multiparty communication-based scenarios. The proposed session key management protocol applies a symmetric polynomial for group members. The polynomial generation method uses security credentials and a secure hash function.

3.1 Experiment setup and results

In this section, we demonstrate the operation of the Probability-Based Keys Sharing protocol we developed, as described in Section 3, with the new approaches to dealing with the vulnerabilities of previous protocols. We performed a lab experiment with 3 Raspberry Pi 3 Model devices (#1: 10.0.0.26, #2: 10.0.0.10, #3: 10.0.0.5). We executed the following steps. **Figure 4** outlines the six stages of generating the whole set of keys and dividing it into subsets with at least one overlapping key, as elaborated herein.

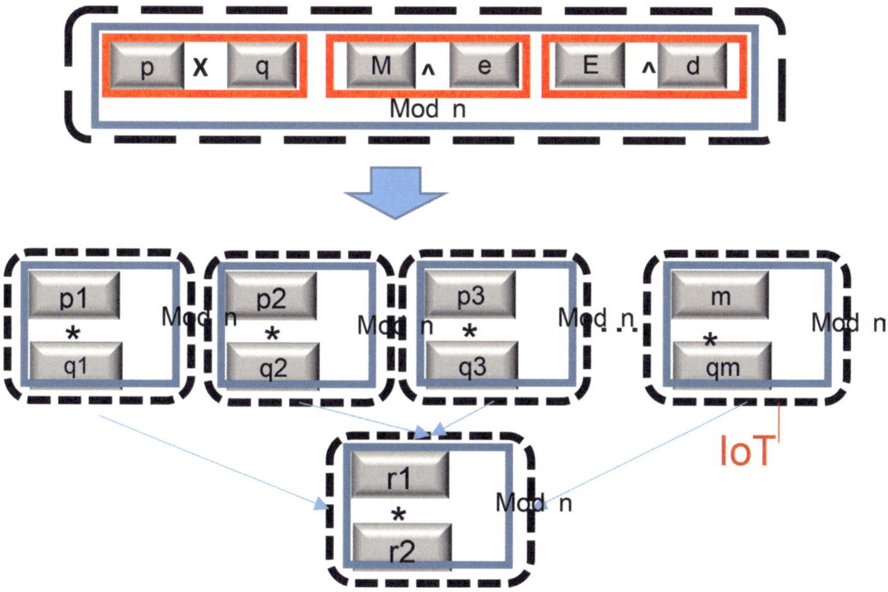

Figure 4.
Splitting and consolidating the modular multiplication into smaller modular multiplication.

Step 1: Preliminary step. The network manager generates a local certificate for each IoT device.

Step 2: Determine who the Primary device is. This operation is done by device #1, which looks for the Primary device on the network, sends a message in broadcast with its certificate to other devices, and declares itself as a master. The other instruments would ignore this message if the device were without a certificate.

Step 3: The Master defines the key-pool size. Device #1 calculates the required pool of keys and the number of keys each device will receive, ensuring an overlap of at least one key between any two devices on this network with a 90% probability. In our case (3 devices), a pool of 152 and 16 keys per device is required.

Step 4: The Master requests the manufacturing of distributed keys. Device #1 (the Master) sends a message in broadcast for all devices to generate keys.

Step 5: Generate distributed keys by each node. Each device generates keys as required. The keys are sent to the Master encrypted with the Master's public key.

Step 6: Distribute the subset keys to each device.

Step 7: Finding a shared key. When a node wants to exchange encrypted messages with another node, it sends a statement with its reference keys. In the experiment, device #2 received a message from device #1 to find a shared key (148).

Step 8: Secure network, Node #2 wants to communicate with node #3 (neither is the Master), and the shared key between them is 42. Node #2 sends a message encrypted by AES using the shared key. Node #3 decrypts it with the same key.

Step 9: Detection of missing devices is employed to discover potential cyberattacks on any network device. The Master (#1) sends each node a ping message every time interval. If there is no answer from a particular device, the current batch of keys is canceled and replaced accordingly.

We executed the experiment step by step, and it went well. We handled several intensive messaging sessions with various key generations and massive messaging.

4. Key sharing using key distribution via a downsized RSA

The most common approach for a secured key distribution is using Asymmetric encryption such as RSA, which executes several modular multiplications for generating the encryption key and for the encryption and decryption stages. A typical IoT device has limited computing resources, which prevents it from executing RSA, compromising key distribution security. In recent related papers, the classic approach replaces RSA with ECC, a similar security level—Saxena and Kawamura [26, 27] proposed parallel processing. Xian-Fu [28] uses GPU achieving a 12% performance improvement. Stergioua [29] execute RSA in Cloud. Goyal [30] recommended ECDH/ECC algorithms. Duy An Ha [5] used ECQV and DTLS, combining authentication and transmission for IoT, and Fadhil and Younis [7] combined multicore CPUs and single-core GPUs.

We propose a downsized RSA implementation in that its results are equivalent to the regular RSA. In this implementation, we split modular multiplications of huge numbers into micro calculations that IOT devices can process. **Figure 4** describes the splitting and consolidating process of a modular N multiplication of two huge numbers, P and Q. Each split component is transmitted to an IoT device to execute it and return the result to the Master device, which consolidates it to provide the required output, encryption key or encrypted/decrypted element.

The processing model comprises the RSA-Distributor and RSA-Observer sub-module.

Figure 5 outlines the RSA-Observer, which is waiting for the RSA-Distributor to send three operads to calculate its modular multiplication, and, when ready, send back its result.

Figure 6 depicts the Distributor's detailed modular multiplication model, from splitting to micro multiplications and distributing it to the available IoT devices for execution. Once all IoT devices' results are accepted, it integrates the detailed results to get the final result, which is then returned as the output of this process.

We conducted an experiment using four connected computers from various manufacturers to prove the model's applicability and assess its effectiveness. The results well support our approach.

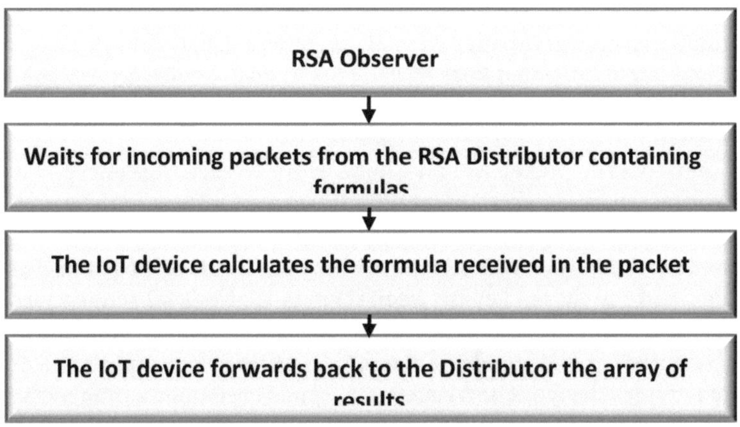

Figure 5.
The RSA-observer.

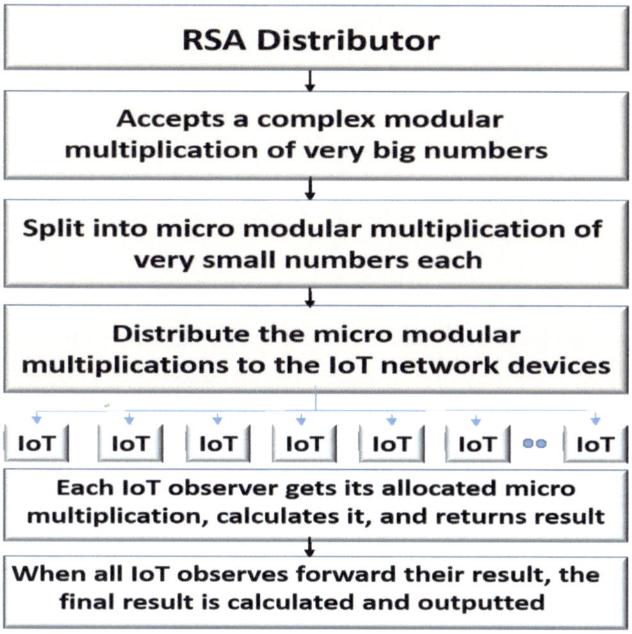

Figure 6.
The RSA-distributor.

5. Key distribution using a reliable internal CA

The increased interconnectivity and interoperability of previously isolated systems have created new attack paths for hidden adversaries. Integrating IoT technologies has led to new attack opportunities for remote adversaries, mainly due to its poor resistance to cyberattacks. For example, the Parking lot attack occurs when the attacker joins the network and accesses hosts in the internal network. In our case, the first stage begins with determining the Master device to serve as the system controller. The Master is selected before executing the key distribution process. The selection of the Master node is exposed to three attacks: a parking lot attack, exposure of the keys dictionary, and a physical attack. A "parking lot attack" is where a malicious device declares itself as the Master and accordingly controls the keys dictionary and the distribution of keys to all devices. We present a complementary solution to this risk by employing existing methods and technologies to protect the distribution protocol against such attacks. We introduce an internal Certification Authority that issues certificates for each IoT device before joining the network. All keys are distributed by the Master to each device using the Unix OS "password" mechanism. If a device "disappears," all encryption keys are immediately replaced.

Eschenauer and Gligor [1] present a selective distribution scheme and revocation of keys to sensor nodes using probabilistic key sharing among the nodes of a random graph. Alagheband and Aref [20] assess KMS for IoT. Sciancalepore [21] focused on avoiding replay attacks and light authentication using ECC.

Our proposal focuses on a local Certificate Authority, a local keywords Dictionary, and a Detector of Missing Devices.

5.1 Local certificate authority method

For security purposes and to neutralize the Parking lot attack, each device holds a certificate on behalf of the local network manager. For this purpose, we use the "Python Own Certificate Authority (OwnCA)" [10], where OwnCA handles the certificates for hosts, servers, or clients. More CAs, such as Certauth 1.3.0, can be found at https://pypi.org/project/certauth. Benslimane and BenAhmed [24] describe the improved protocol:

1. A preliminary step was added to the protocol, where the local network administrator provides a local certificate for each device connected to the network.

2. All messages exchanged between devices include their certificate to prevent a malicious device from being added to the network.

5.2 Own keywords dictionary method

An IoT device in a local network may be attacked physically, leading to key discovery and network hackery. Therefore, the Primary device sends a mechanism to secure the keys to prevent such incidents. The system operates similarly to the agent that ensures the password and shadow files in the UNIX system. The *keyword* file contains the following data: Device name, Key index number – encrypted data (SHA256), Keys – Encrypted data (SHA256), Last key change time – visible data, and Key Expiration Time – Visible data.

5.3 Detecting missing devices

A daemon is activated in the Master device that checks all active devices to prevent a malicious opponent from "snatching" a device and extracting the information required to hack the network. The Master pings each device to reveal those who do not reply within seconds. After three unanswered tries, the Master eliminates these unanswered devices and changes the keys for all remaining devices in the network.

6. Conclusions

This chapter deals with security issues explicitly raised in IoT networks due to their limited resources and capacity. The conventional way to deal with security issues related to network inter-node messaging is to encrypt the data passing through the network using Symmetric encryption, which requires frequent key replacement and a distribution system. We presented various key distribution methods from the literature and described their operation principles. We elaborated on three advanced techniques our team designed, developed, and experimented them: (i) probability-based Key Sharing exploiting the available IoT computing capabilities to generate keys transmitted to the Master and redistribute them to the nodes ensuring the existence of at least one shared key between any two nodes that may need to interconnect. (ii) Lightweight RSA key delivery utilizing free IoT capacity to execute low-scale modular multiplications required for RSA-secured key distribution. (iii) Internal CA key generation and local distribution. These three methods follow the concept of having the IoT network members internally handle the entire security measurements without needing external servers. Methods (i) and (ii) use the excess computing capacity of the devices in the local network resulting in a win-win situation. Method (iii) requires adding a dedicated machine as the internal CA. Although we presented various solutions to the security challenges in networks characterized by low capabilities, it also applies to any network. It may be a better solution in cases with similar attributes since security issues in IoT and other networks are still amidst cyber security interest and research. To conclude this chapter, we may say that security challenges are still ahead of us; we should first strive for solutions that exploit existing capacity without forcing the embedding of new technologies foreign to the existing setup.

Author details

Menachem Domb
Ashkelon Academy College [AAC], Ashkelon, Israel

*Address all correspondence to: dombmnc@edu.aac.ac.il

References

[1] Eschenauer L, Gligor VD. A key-management scheme for distributed sensor networksnetworks. In: Proceedings of the 9th ACM Conference on Computer and Communications security (CCS '02). USA, NY, Washington DC: ACM; 11 2002. pp. 41-47. DOI: 10.1145/586110.586117

[2] AbuAlghanam O, Qatawneh M, Almobaideen W, Saadeh M. A layered architecture and protocol for key distribution in the context of IoT in smart cities. Journal of Information Security and Applications. 2022;**67**:103173. DOI: 10.1016/j.jisa.2022.103173

[3] Hamid MA, Wadud MA-A, Hassan MM. A key distribution scheme for secure communication in acoustic sensor networks. Future Generation Computer Systems. 2018;**86**:1209-1217

[4] Naoui S et al. Security analysis of existing IoT key management protocols. In: 2016 IEEE/ACS 13th International Conference of Computer Systems and Applications (AICCSA). NY, USA: IEEE; 2016

[5] Metan J, Murthy KNN. N-tier modelling of robust key management for secure data aggregation in wireless sensor network. International Journal of Electrical and Computer Engineering (IJECE). 2019;**9**:2682-2690

[6] Veena RS et al. A cost-effective 2-tier security paradigm to safeguard cloud data with faster authentication. International Journal of Electrical and Computer Engineering (IJECE). 2019;**9**:3833-3842

[7] Fadhil HM, Younis MI. Parallelizing RSA algorithm on multicore CPU and GPU. International Journal of Computer Applications (0975-8887). (Berlin, Germany: Researchgate). 2014;**87**(6):1-8

[8] Goyal TK, Sahula V. Lightweight security algorithm for low power IoT devices. In: International Conference on Advances in Computing, Communications and Informatics (ICACCI), Jaipur, India. Piscataway, New Jersey, United States: IEEExplore. 2016. pp. 1725-1729. DOI: 10.1109/ICACCI.2016.7732296

[9] Usman M, Ahmed I, Imran Aslam M, Khan S, Usman UASM. A lightweight encryption algorithm for securing IoT devices. International Journal of Advanced Computer Science and Applications. (The Science and Information (SAI) Organization, Queens, New York. Operated from offices in the United States, the United Kingdom, and India). 2017;**8**(1):1-10

[10] Nandhini P, Vanitha V. A study of lightweight cryptographic algorithms for IoT. International Journal of Innovations & Advancement in Computer Science. (SEMANTIC SCHOLAR, North Northlake Way, Suite 110, Seattle, WA 98103). 2017;**6**(1):1-7

[11] Iqbal U, Bhola J, Jayasudha M, Ahmad MW, Netware R, Yadav AR, et al. ECC-based authenticated key exchange protocol for fog-based IoT networks. Security and Communication Networks. 2022, 2022:15. DOI: 10.1155/2022/7264803

[12] Eldefrawy MH, Pereira N, Gidlund M. Key distribution protocol for industrial internet of things without implicit certificates. IEEE Internet of Things Journal. 2019;**6**(1):906-917. DOI: 10.1109/JIOT.2018.2865212

[13] Lian H, Yang Y, Zhao Y. Efficient and strong symmetric password authenticated key exchange with identity privacy for IoT. In: IEEE Internet of Things Journal. NY, USA: IEEE; 2022. DOI: 10.1109/JIOT.2022.3219524

[14] Salman O, Abdallah S, Elhaji IH, Chehab A, Kayssi A. Identity-based authentication scheme for internet of Things. In: IEEE Symposium on Computers and Communication (ISCC). Messina, Italy, NY, USA: IEEE; 2016. pp. 1109-1111. DOI: 10.1109/ISCC.2016.7543884

[15] Shivraj VL, Rajan MA, Singh M, Balamuralidhar P. One time password authentication scheme based on elliptic curves for internet of things (IoT). In: The 5th IEEE National Symposium on Information Technology: Towards New Smart World. NY, USA: IEEE; 2016. 2015. pp. 1-6

[16] Aman MN, Taneja S, Sikdar B, Chua KC, Alioto M. Token-based security for iot with dynamic energy-quality tradeoff. IEEE Internet of Things, Journal. (NY, USA: IEEE: 2016). 2018

[17] Guo H, Zheng Y, Li X, Li Z, Xia C. Self-healing group key distribution protocol in wireless sensor networks for secure IoT communications. Future Generation Computer Systems. 2018;**89**:713-721. DOI: 10.1016/j.future.2018.07.009

[18] Moharana SR, Jha VK, Satpathy A, Addya SK, Turuk AK, Majhi B. Secure key distribution in IoT cloud networks. In: 2017 Third International Conference on Sensing, Signal Processing and Security (ICS). NY, USA: IEEE; 4 May 2017. pp. 197-202

[19] Othman W, Fuyou M, Xue K, Hawbani A. Physically secure lightweight and privacy-preserving message authentication protocol for VANET in Smart City. IEEE Transactions on Vehicular Technology. 2021;**70**(12):12902-12917. DOI: 10.1109/TVT.2021.3121449

[20] Alagheband, Mahdi R, Aref MR. Dynamic and Secure Key Management Model for Hierarchical Heterogeneous Sensor Networks. UK, Londom: IET Information Security 6.4; 2012. pp. 271-280. DOI: 10.1049/ietifs.2012.0144

[21] Sciancalepore, Piro G, Boggia G, Bianchi G, Capossele A. Key management protocol with implicit certificates for IoT systems. In: Savio Proceedings of the Workshop on IoT Challenges in Mobile and Industrial Systems. Florence, Italy, NY, USA: ACM; 2015. pp. 37-42. ISBN: 9781450335027

[22] Roman R, Alcaraz C, Lopez J, Sklavos N. Key management systems for sensor networks in the context of the internet of things. Computers & Electrical Engineering. 2011;**37**(2):147-159

[23] Wazid M, Das AK, Odelu V. Design of secure user authenticated key management protocol for generic IoT networks. IEEE Internet of Things Journal. 2017, 2018;**5**(1):269-282

[24] Benslimane KBA. Efficient end-to-end secure key management protocol for IoT. International Journal of Electrical and Computer Engineering (IJECE). 2017;7(6):3622-3631

[25] Mahmood Z, Ning H, Ghafoor A. A polynomial subset-based efficient multiparty key management system for lightweight device networks. Sensors. 2017;**17**(4):670. DOI: 10.3390/s17040670

[26] Saxena S, Kapoor B. State of the art parallel approaches for RSA public key base cryptosystem. International

Journal on Computational Sciences & Applications (IJCSA). (USA: Cornell University). 2015;**5**(1):81-88. DOI: 10.48550/arXiv.1503.03593

[27] Kawamura S, Koike M, Sano F, Shimbo A. Parallel computation of the generating keys for RSA cryptosystems. Electronics Letters;**32**(15):1365-1366, IEEE Explorer

[28] Xian-FuWong B-MG, Lee W-K, Phan RC-W. Performance evaluation of RSA and NTRU over GPU with Maxwell and Pascal architecture. Journal of Software Networking. 2017;**2017**(1):201-220

[29] Christos Stergioua, Kostas E. Psannisa, Byung-GyuKimb, Brij Guptac, Secure integration of IoT and cloud computing, Elsevier, Future Generation Computer Systems, 78, Part 3, 2018, 964-975

[30] Goyal TK, Sahula V. Lightweight security algorithm for low power IoT devices. In: 11-2016 Conference on Advances in Computing, Communications, and Informatics (ICACCI). NY, USA: IEEE; 2026. DOI: 10.1109/ICACCI.2016.7732296

Chapter 8

Artificial Neural Network Computational Techniques in Biometric Handwriting

Jose Luis Vásquez-Vasquez and Carlos M. Travieso-González

Abstract

This study presents a novel methodology that combines the power of multilayer perceptron (MLP) neural networks with validated graphometry approaches for individual identification based on handwriting. By integrating the computational capabilities of MLPs with the graphometry characteristics utilized in graphology, this proposal aims to leverage the distinctiveness and stability of both approaches. Handwriting, as a widely accepted behavioral biometric characteristic, serves as a reflection of an individual's personality, enabling effective identification. The MLP's ability to learn complex relationships between inputs and outputs, coupled with the graphometry measures capturing intricate patterns within the data, contributes to developing highly accurate and efficient identification systems. This comprehensive approach fuses the strengths of MLP neural networks and graphometry techniques, providing a promising avenue for advancing the field of personal identification through handwriting analysis. By harnessing the intrinsic uniqueness of handwriting and its equivalence to other behavioral traits, the methodology enables discerning a person's psychological profile and overcomes variations over time. The implementation of identification systems based on these properties establishes robust and reliable solutions in personal identification.

Keywords: handwriting, graphometry, identification, computational techniques, behavioral biometric

1. Introduction

Handwriting, as a biometric-behavioral trait, operates on a deeply subconscious level, offering substantial and reliable information for person recognition [1, 2]. While the signature serves as a widely accepted means of legal authentication, other aspects of a person's handwriting can also be analyzed for identification purposes, such as in forensic studies that determine or certify document authorship—a task typically entrusted to handwriting experts.

The act of writing encompasses a complex phenomenon influenced by various personal factors. Like how an individual can be recognized by their unique laughter, gestures, or gait, their writing style is shaped by their personality and physiological traits [3] endowing it with identification potential.

In line with other pattern recognition domains, research on handwriting for biometric identification revolves around three key aspects: feature extraction for effective representation of the phenomenon, utilization of encoding methods and models, and selection of the most suitable classifier that aligns with the recognition task and the properties of the developed model. Studies in this field explore diverse approaches to feature extraction. Some focus on contour-based features [4] or employ bi-quadratic interpolation to capture letter curvature information [5]. Others propose feature extraction through wavelet transforms of handwriting images [6] or Hermite coefficients derived from text lines [7].

In addition to feature extraction, research in this field encompasses the study of different types of classifiers commonly used for handwriting-based identification. Artificial Neural Networks (NN) [8, 9], K-Nearest Neighbors (KNN) [10], Support Vector Machine (SVM) [11, 12], Hidden Markov Models (HMM) [13], and Gaussian Mixture Models (GMM) [14, 15] are some examples commonly used in this field. These classifiers play a crucial role in achieving accurate identification outcomes.

A strong line of research, on which much of this work is based, focuses on the extraction and use of structural features in handwriting [16, 17]. Structural or graphometry features are commonly employed in forensic document analysis by handwriting experts and are also utilized in graphology to determine a person's psychological profile.

Handwriting characteristics can be classified into two main groups: those that are quantitatively evaluated by measuring traces, and those that require a more qualitative analysis by an expert. Quantitative features include the size of ascenders and descenders, inclination angles, aspect ratios, proportionality index, and the size of the calligraphic box. On the other hand, qualitative features relate to the richness of gestures and particularities in strokes, such as the way certain letters begin or end in a word, the roundness of letter forms, and the overall legibility of the writing. Evaluating these qualitative characteristics poses a challenge in automated processing and remains an open research topic.

Research in this field has primarily focused on solving two problems: word or handwritten text recognition (RATM) and writer recognition [18]. Significant advancements have been made in achieving high success rates in both areas, reflecting the progress of dedicated research efforts. The utilization of diverse classifiers and the exploration of structural and graphometry features contribute to the development of robust and effective handwriting identification systems.

These endeavors contribute to advancing the field of handwriting-based biometric identification, enabling more accurate and sophisticated systems for person recognition.

1.1 Motivation

The analysis of handwriting, whether conducted manually or automatically, presents a significant challenge due to the inherent variability observed among samples from the same writer. This variability can stem from various factors, including the specific conditions in which the act of writing occurs and the changes that naturally occur over time. For instance, individuals using banking services often find themselves repeatedly asked to replicate their signature as it appears on their identity documents. However, this self-imitation can become indistinguishable from forgery, particularly in the case of older individuals whose handwriting changes become more pronounced with age.

The situation arises from a lack of attention to the changes in handwriting associated with the aging process. Interestingly, if these changes were considered, they could potentially provide more reliable means of identification than a mere copy of an original document. Despite the numerous challenges involved in handwriting analysis, significant progress has been made in its automatic processing in recent years [19, 20]. However, when it comes to automatic writer recognition, the writing samples commonly used in research are often obtained within short time periods. As a result, they fail to accurately capture the evolving characteristics of handwriting caused by the aging process. In fact, it can be argued that existing research does not delve deeply enough into this issue, leading to a lack of comprehensive analysis regarding this phenomenon. Given the increasing social and economic engagement of older populations, addressing this issue becomes even more crucial. It is imperative to investigate and account for the long-term variations in handwriting to develop robust automated writer recognition systems that better reflect the reality of handwriting changes over time.

The successful recognition of individuals through their handwriting, despite changes over time, requires a comprehensive investigation of handwriting characteristics and advanced processing techniques. This research focuses on two key aspects: the development of robust computational techniques capable of handling data variability, and the analysis of techniques and characteristics employed by handwriting experts.

The study of handwriting characteristics is particularly relevant in forensic analyses, where the comparison of documents written by the same person at different times becomes necessary. Addressing the challenges posed by potential changes in handwriting is a primary objective of this work. By identifying consistent graphical elements, we aim to enhance the writer recognition process, thereby increasing the reliability and security of systems while minimizing user inconveniences.

By synergizing robust computational techniques with insights from handwriting expertise, this research aims to contribute to the advancement of highly reliable and accurate writer recognition systems. The objective is to ensure the effectiveness of these systems, even in the presence of natural variations in handwriting that occur over time. Through this interdisciplinary approach, we strive to enable robust identification methods that can withstand the challenges posed by handwriting variability, enhancing the overall performance and applicability of such systems.

1.2 Relative works

Currently, the term "Biometrics" [21] is used to refer to the technological field dedicated to the identification of individuals based on their physiological or behavioral biometric traits, e.g., fingerprints, iris, handwriting, hand geometry and others [22].

Physiological biometric traits include iris, fingerprint, hand geometry, retina, and DNA. Their main quality is that they have little or no variability over time, but their acquisition is more invasive and requires the cooperation of the subjects, while behavioral or behavioral biometric traits, such as voice, signature or handwriting in general, are less invasive although the accuracy of identification is lower due to the variability of behavioral patterns.

A personal trait will be valid, and a biometric system will be able to distinguish people based on it if it fulfills the following properties:

- Universality: every person must possess such a biometric trait.

- Uniqueness: different people must possess different traits, sufficiently different to allow them to be distinguished based on that trait.
- Permanence: the trait must be sufficiently invariant over time.
- Measurability: the trait must be quantitatively characterizable.

Writing, as a system of graphic representation of language, holds the potential to be a distinctive biometric trait. It involves the creation of engraved or drawn signs on a flat surface, with paper currently being the most widely used medium, closely followed by digital devices.

The process of writing is learned and shares common techniques among individuals who speak the same language. Initially, it is a conscious and deliberate act, but with time and practice, it becomes ingrained in our subconscious, becoming a reflexive action. This transition from volitional to reflexive writing explains the enduring and unique graphic characteristics that are specific to everyone, allowing for their differentiation and distinction from others.

Identifying individuals through their handwriting is of utmost importance and finds applications in various domains [23]. In the realm of forensics, handwriting identification plays a critical role in determining document authorship and aiding in the resolution of legal cases.

Moreover, the ability to identify individuals based on their handwriting is also highly relevant in the context of security and access control [24]. Handwriting recognition systems are deployed in settings where ensuring the accurate identification of individuals is crucial, such as financial institutions, airports, and cutting-edge security systems.

Continual research and development in the field of biometric identification have propelled the study of identifying individuals through their handwriting, making it a topic of enduring interest and ongoing advancement. Researchers are exploring innovative techniques, including the analysis of structural and dynamic features of handwriting, as well as leveraging machine learning algorithms [25] and neural networks [26] to enhance the precision and reliability of handwriting identification systems. The evolving landscape of handwriting analysis presents opportunities to improve the accuracy and efficiency of biometric identification based on handwriting. This dynamic field continues to push the boundaries, enabling advancements in the science of identifying individuals through their unique writing characteristics.

In [27], the authors present a proof-of-concept for a cognitive-based authentication system that utilizes an individual's writing style as a unique identifier to grant access to a system. They train a machine learning SVM model on stylometric features to effectively distinguish between texts generated by different users. The extracted stylometric feature vector is then used as input to a key derivation function, generating a unique cryptographic key for each user. Experimental results demonstrate the system's accuracy of up to 87.42% in classifying texts as written and validate the security and uniqueness of the generated keys. This research explores the intersection of natural intelligence, cognitive science, and cryptography, aiming to develop a cognitive cryptography system. By leveraging behavioral features from linguistic-biometric data through stylometry, the proposed system detects and classifies users, generating cryptographic keys for authentication and enhancing access control security.

2. Material

The database used in this study was created at the University of las Palmas de Gran Canaria and contains handwriting samples from 100 different people, each of whom made 10 handwritten copies of the following text in Spanish (**Figure 1**).

From the proposed text, 34 words were selected whose images were extracted from each of the previously scanned writings. The database contains a greyscale image of each of the words, for a total of 34,000 images (**Figure 2**).

The main conditions for the elaboration of the writings used to form the database are the following:

- All writers copied the same text and used the same type of pen.

- All samples were written on 80 gram DIN-A4 paper, using a flat, rigid surface as a support.

- The copies were made over the course of a week, on different days and at different times, depending on the availability of each writer.

In 1605 the first edition of "El Ingenioso Hidalgo Don Quijote de la Mancha", the most universal work of Spanish literature, written by Cervantes, was published.

It is a perfect caricature of chivalric literature, and its main characters, Don Quixote and Sancho Panza, embody the two types of the Spanish soul, the idealist and dreamer, who forgets the necessities of material life to run after inaccessible chimeras, and the positivist and practical, though rather fatalistic.

This prized jewel of Spanish literature has managed to conquer the whole world, and is perhaps, with the Bible, the work that has been translated into more languages than any other, its characters becoming true archetypes of universal category.

Figure 1.
Text used to formalize the writers' database.

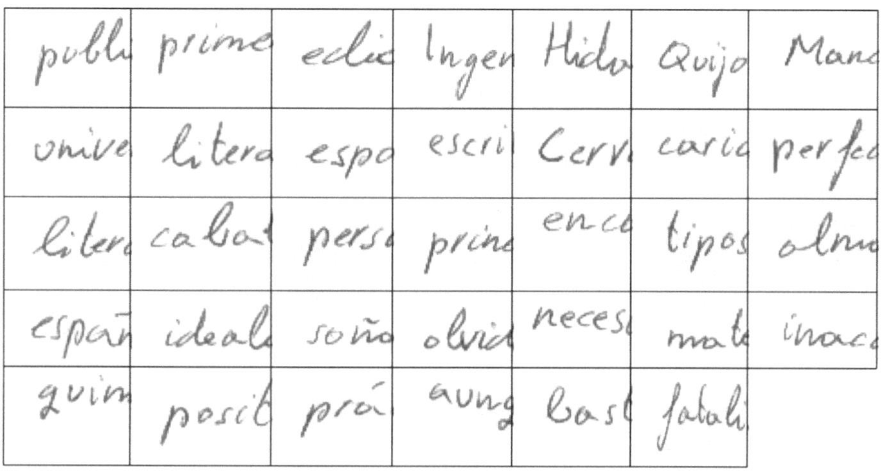

Figure 2.
Words extracted.

3. Methods

3.1 Methodology

The proposed methodology adopts a classical structure of pattern recognition systems configured in identification mode, which avoids the need for conducting multiple experiments with different evaluation modes. In this mode, the objective is to accurately assign each sample to its corresponding class. The methodology consists of several phases: first, all images undergo pre-processing to ensure optimal conditioning. Subsequently, various processing techniques are applied to extract the features described in the previous chapter.

The subsequent stages of the experimentation focus on applying supervised classification in the identification mode. Three classifiers, namely KNN, NN, and SVM, are utilized for the calligraphic features, while a combination of HMM and SVM is employed for contour characterization. The use of multiple classifiers for the calligraphic parameters aims to determine whether the obtained results are primarily attributed to the effectiveness of the parameters rather than the idiosyncrasies of a specific classifier. This approach enhances the reliability and comprehensiveness of the experimental analysis, providing a robust assessment of the proposed methodology.

After parameterizing the contour processing, the generated models with HMM were subjected to transformation using Fisher's Kernel. This transformation mapped the models to hyperdimensional spaces, which served as input data for a classification system based on SVMs. The choice of SVMs is supported by the favorable results obtained in previous research studies, such as those conducted [28–31]. These studies utilized large datasets, highlighting the effectiveness of SVMs in handling substantial amounts of data. By applying SVMs in the classification system, the methodology leverages their proven performance to achieve accurate and reliable results in the analysis of contour data.

The system modeling technique in this study involved two phases: training and testing. To ensure reliable results, the databases were divided into two groups, with each group used for a specific phase. The Hold-out cross-validation technique was employed for the division of the groups. This technique ensures that the systems are trained and tested on entirely different samples, reducing the risk of bias.

To further enhance the reliability of the results and avoid dependence on specific samples, a significant number of iterations were performed. This statistical independence was achieved by repeatedly applying the cross-validation method, dividing the dataset into training and test subsets.

Moreover, to assess the robustness of the systems, the percentages of training and test samples were varied, ranging from 50% to 20%. This variation allowed for a comprehensive analysis of the system's performance under different proportions of data.

Additionally, singular experiments were conducted to evaluate the stability, invariance, and robustness of the systems in more complex scenarios. These experiments aimed to test the system's ability to handle diverse and challenging situations, providing insights into its performance under different conditions.

3.1.1 Support Vector Machine

The Support Vector Machine (SVM) is a learning system that has undergone significant development in recent years, both in the generation of new algorithms and in the strategies for their implementation. It is used to train linear learning machines efficiently, both for classification and regression (linear and non-linear). Formally, a Support Vector Machine is a static network based on kernels that performs classification on vectors transformed to a higher dimensional space, separated by a hyperplane in the transformed space, which is originated by the kernel itself. The basic operations performed by an SVM are listed below:

- Transform the data to a higher dimensional space through a previously defined kernel function. This reformulates the problem by implicitly mapping the data to the new space.

- Find the hyperplane that maximizes the margin between the nearest training class patterns by performing an efficient computation of the optimal hyperplane.

- If the data are not linearly separable find the hyperplane that maximizes the margin and minimizes a function of the number of misclassifications.

3.1.2 Hidden Markov Models

Hidden Markov Models first appeared in recognition systems in the late 1970s, allowing the development of specific probabilistic techniques for the estimation of these systems. Today, they are very efficient, robust and computationally flexible techniques for many types of recognition systems.

The main difference between an HMM and a Markov chain is that each state in a Markov chain is deterministically associated with a single output observation value, whereas in an HMM each is associated with a probability distribution of all possible output observation values.

An HMM can be viewed as a finite state machine in which two well-defined stochastic processes interact with each other: one of them remains unobservable and acts in the background (it is the hidden layer of the model) behind another observable that produces the sequence of output observations. The former involves a set of states connected to each other by means of transitions with probabilities; while the latter consists of a set of output observations, each of which can be emitted by any of the states according to a probability function associated to each of them. In other words, an HMM consists of a Markov chain and a set of probability functions associated with each state; that is, states are no longer simply symbols but are now associated with sets of probability distributions. Transitions between states depend on the occurrence of some symbol.

3.2 Training phase

In this phase, we determine the values of the optimization variables that characterize each of the shape classes to be classified. These variables include the number of neurons in the hidden layer and the number of networks.

When using neural networks for classification, the optimal number of neurons in the hidden layer is determined through a trial-and-error process. We explore different configurations to find the one that yields the best performance.

Furthermore, in most cases where neural networks are employed as classifiers, multiple networks are utilized in parallel. This approach ensures that the classification result is independent of the initial values of the model. The final decision for classifying a particular sample follows the "most voted" strategy, considering the opinions of the parallel neural networks. The experimentation process also includes determining the appropriate number of parallel neural networks for each scenario, striking a balance between performance and efficiency.

When utilizing neural networks for classification, the optimal number of neurons in the hidden layer was determined through an iterative trial and error process. Various configurations were tested to find the most suitable number that yielded optimal performance and accuracy.

Moreover, in most cases where neural networks were employed as classifiers, multiple networks were employed simultaneously in a parallel manner. This approach aimed to mitigate the impact of initial values on the model and enhance the robustness of the classification results. By adopting a "most voted" strategy, the final classification decision for a particular sample was based on the collective input from the parallel networks. The experimentation process also involved exploring the appropriate number of parallel neural networks for each scenario, ensuring reliable and consistent classification outcomes.

3.2.1 Number of neurons in the hidden layer and number of networks

The optimal number of neurons in the hidden layer for classification using neural networks was determined through an iterative trial and error process.

Furthermore, to enhance the reliability and stability of the classification results, multiple neural networks were often employed in parallel as classifiers. This approach aimed to reduce the impact of initial values on the model and increase the robustness of the overall system. In this scheme, the final classification decision for a specific sample was determined based on a "most voted" strategy, where each parallel network contributed its prediction, and the class with the highest number of votes was selected.

The experimentation process also involved exploring the appropriate number of parallel neural networks for each situation. This step allowed for the identification of the optimal configuration that maximized the accuracy and consistency of the classification outcomes.

3.2.2 SVM parameters

The version used in this work is the SVM^{light} with radial basis kernel function (RBF), defined by [32]:

$$K(x,y) = e^{-\gamma\|x-y\|^2} = e^{\frac{\|x-y\|^2}{2\sigma^2}} \quad (1)$$

In this configuration, the crucial parameter to determine is the variance (σ^2) of the Gaussian (gamma (γ)), which is inversely proportional to the variance of the Gaussian kernel and represents the width of the RBF kernel.

Regarding the SVM, another important parameter to define is the training cost constant (c). Selecting an appropriate value for this parameter in real-world applications can be more challenging than choosing the kernel. The optimal value of 'c' often depends on the nature of the data. In all experiments conducted, a value of 10 was used for 'c'.

3.2.3 HMM state numbers

In the contour characterization using HMM, the determination of the number of states is a crucial parameter during the system modeling phase. It directly impacts the reliability, robustness, and stability of the system. To determine the optimal number of states, multiple experiments were conducted using different values. Based on the classification results obtained from these experiments, the most appropriate number of states was selected. This careful selection ensures that the system achieves the desired performance and accurately captures the patterns and characteristics of the contour data.

3.3 Graphometry feature extraction

As stated in the introductory section, graphometry characteristics play a pivotal role in the forensic analysis of documents, as they encompass a substantial number of parameters utilized by handwriting experts. In alignment with the research objectives, we conducted an extensive review of these measures. We considered factors such as the significance of each characteristic in defining an individual's handwriting style and its effectiveness in discerning between different writers. Moreover, we considered the computational complexity involved in automatically extracting these features, which is a fundamental aspect to consider. By taking these factors into consideration, we aimed to ensure that the selected morphometric measures are both relevant and practical for our research purposes.

3.3.1 Length of ascenders and descenders

"Ascenders" and "descenders," commonly referred to as "hampas" and "jambs" in the field of forensic document analysis, are key features extensively examined by handwriting experts. Ascenders represent the upper extensions of letters, while descenders refer to the lower extensions (refer to **Figure 3** for visual illustration). These features hold significant importance in the analysis of documents, enabling experts to gather valuable insights about the handwriting style in question. By studying the hampas and jambs, experts can gain a deeper understanding of the writer's unique characteristics and tendencies, aiding in the identification and comparison of different handwriting samples.

The significance of these features in the biometric study of writing becomes evident when we consider their prevalence in the Latin alphabet. Ascenders and descenders are commonly found in approximately 50% of the letters within this alphabet. This statistic, as illustrated in **Table 1**, highlights the widespread presence of these features in written text. By considering their frequency, handwriting experts can effectively utilize ascenders and descenders as reliable indicators and

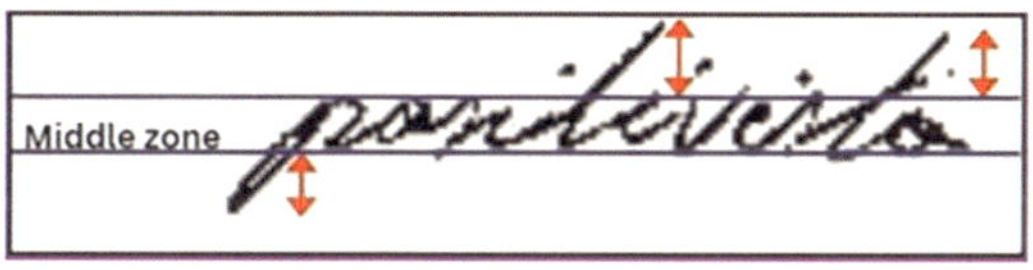

Figure 3.
"Ascending and descending".

Letter	Italics		Print	
	Ascenders	Descenders	Ascenders	Descenders
b	√	—	√	—
d	√	—	√	—
f	√	√	√	—
g	—	√	—	√
h	√	—	√	—
j	—	√	—	√
k	√	—	√*	—
l	√	—	√	—
p	—	√	—	√
q	—	√	—	√
t	√	—	√	—
y	—	√	—	√
z	—	√	—	—

Table 1.
Presence of ascenders and descenders in Latin albatross letters.

discriminators in the analysis of documents. These features provide valuable information about a writer's style, aiding in the identification and differentiation of various handwriting samples.

A classical method for extracting ascenders and descenders is to divide the text into three zones: the upper zone, the middle zone, and the lower zone. However, determining these zones is not an easy task, the most common method in image processing being the horizontal projection, see **Figure 4**.

Various algorithms have been proposed to determine the boundaries of each zone based on the horizontal projection [33, 34]. However, it is important to note that no single method can be universally applied to all cases. For instance, Kirli and Gülmezoğlu [34] suggest minimizing a specific expression to calculate the base and top lines of a text. This demonstrates the diversity of approaches used to address this challenge in handwriting analysis. Researchers continuously strive to develop effective methods tailored to different scenarios, considering the specific characteristics and requirements of each case.

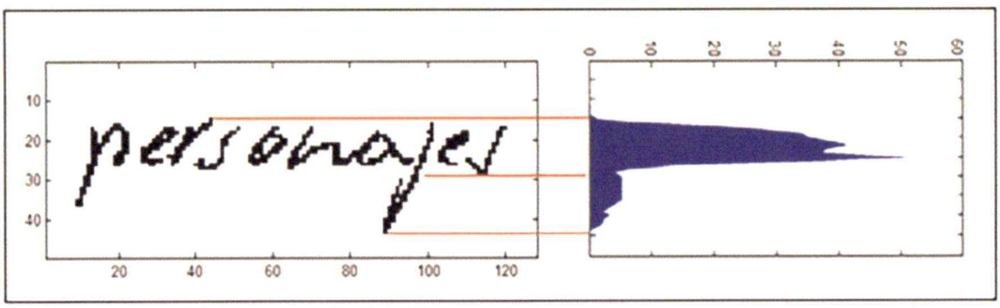

Figure 4.
Horizontal projection to determine the body of a word.

$$D(L_2, L_3) = \sum_{L=L_1}^{L_2-1} (P_{min} - P_L)^2 + \sum_{L=L_2}^{L_3} (P_{max} - P_L)^2 \quad (2)$$

$$+ \sum_{L=L_3-1}^{L_4} (P_{min} - P_L)^2 \quad (3)$$

The upper line of the word boundary is denoted by L_1, while the lower line is represented by L_4. L_2 and L_3 correspond to the upper and baseline, respectively. The minimum and maximum values of the projection are represented by P_{min} and P_{max}. The value PL corresponds to the projection index L. To determine the optimal positions of L_2 and L_3, the total squared error, denoted as $D(L_2,L_3)$, is minimized. The positions that yield the minimum value of $D(L_2,L_3)$ correspond to the final placement of the base and top lines.

Once the boundaries of the word body have been established, the next step involves identifying the ascending and descending strokes within the text. Additionally, their lengths are measured relative to the total height of the word. This analysis provides valuable insights into the proportions and characteristics of the handwriting, contributing to the overall understanding of the writer's style (**Figure 5**).

$$f_1 = \frac{\text{Hampa length}}{\text{Total height}} \quad (4)$$

$$f_2 = \frac{\text{Jamb length}}{\text{Total height}} \quad (5)$$

3.3.2 Skew

Skew refers to the inclination of words in relation to the x-axis of the Cartesian system. However, there is some debate among authors regarding its specific definition. Some argue that skew pertains to the inclination of the entire line of writing with respect to the x-axis.

In handwriting recognition applications, skew is typically detected and corrected during the pre-processing stage. This is because skew, as a characteristic, is influenced by various factors, including the writer's state of mind, and does not provide relevant information for recognition. Instead, it can hinder the recognition process. However, in the context of a biometric system, handwriting experts have observed that everyone tends to write with a relatively consistent skew, making it a commonly analyzed characteristic.

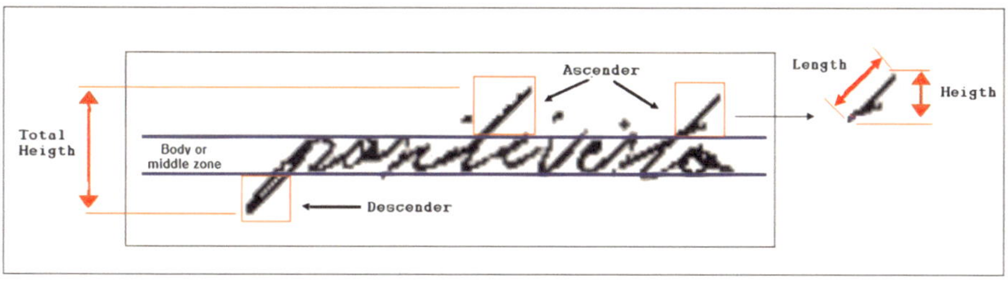

Figure 5.
Length of jambs and jambs as a percentage of the total height of the word.

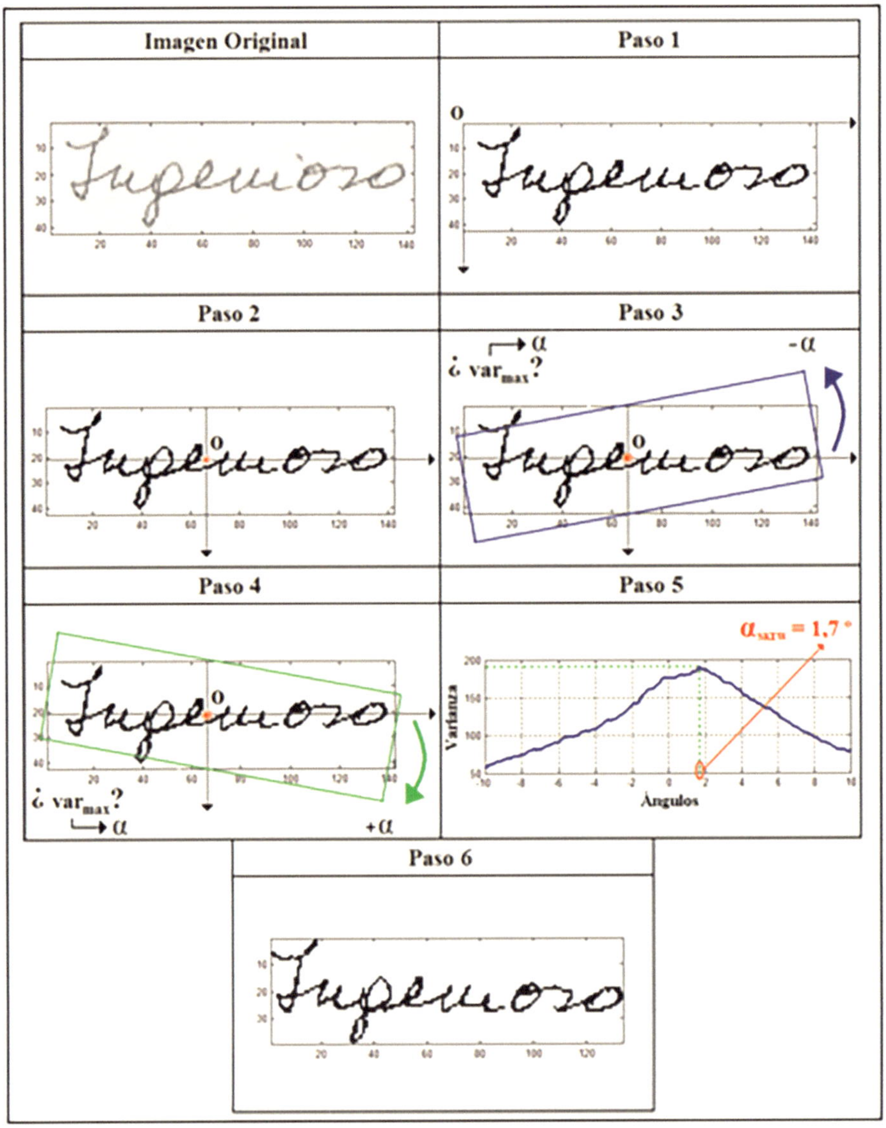

Figure 6.
Example of a breakdown of the skew detection and correction method.

Considering that skew is a variable dependent solely on the writer, it was decided to estimate it as the mean value of the processed words. However, it is important to note that this estimation may not accurately represent the true value of the parameter.

The method developed for the quantification of this parameter is based on horizontal projection techniques with a series of modifications described below. In summary, the process consists of 6 steps:

1. Starting from the original word image (in greyscale) binarisation is performed using Otsu's method [35].

2. The centre of mass is calculated from the obtained image, which will be used as a reference point of rotation, which allows a better correction for words with an oscillating skew or sinusoid. Next, the range of angles used for rotation is defined, which in this project goes from −10 to 10 degrees in steps of 0.1 degrees ($\alpha \in [-10{:}0.1{:}10]$).

3. In this step, a rotation of the image with respect to the centre of mass is performed with each of the negative angles. On each of these images, a horizontal projection is performed where the cost function does not look for the maximum value of the foreground pixels per row, but for the maximum variation of these pixels.

4. This step is the same as the third one, but here the image is rotated with each of the positive angles, also looking for the maximum variance of the foreground pixels per row.

5. With all the variances obtained for each angle of rotation, we proceed to evaluate for which of them the maximum variance is obtained. This will be the angle of inclination (αSKEW) of the word with respect to the 'x' axis.

6. Here a rotation of the image is performed with the opposite angle of the tilt angle obtained in the previous step. **Figure 6** shows each of the steps listed, for a concrete example of a database image:

3.3.3 Slant

Slant refers to the deviation from the vertical or 'y' axis of the Cartesian system exhibited by each letter within a word. It stands out as one of the most prominent characteristics of an individual's writing style, making it a significant parameter for quantification by handwriting experts.

While the study of slant typically focuses on analyzing individual letters, in our case, it was extracted from words. When referring to the slant of a line, we describe it as upright or vertical when the line's axis forms a 90-degree angle with the base of the line. Any deviation from this vertical position indicates an inclination. In our analysis, if the inclination is towards the right of the vertical, the angle of inclination is considered positive, whereas if it leans towards the left, it is considered negative. The vertical axis (90°) serves as the reference point with 0° representing no inclination. For visual reference, please refer to **Figures** 7 and **8**.

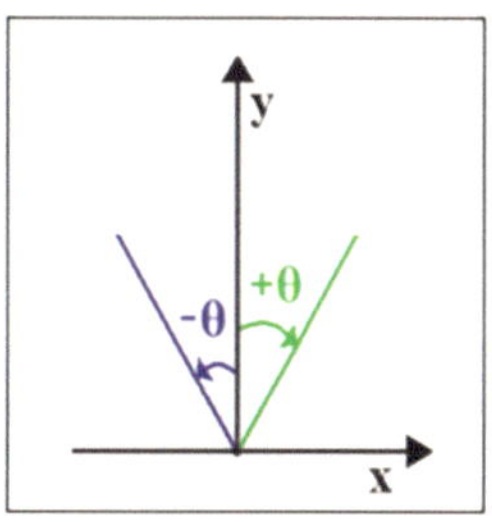

Figure 7.
Reference system chosen for slant estimation.

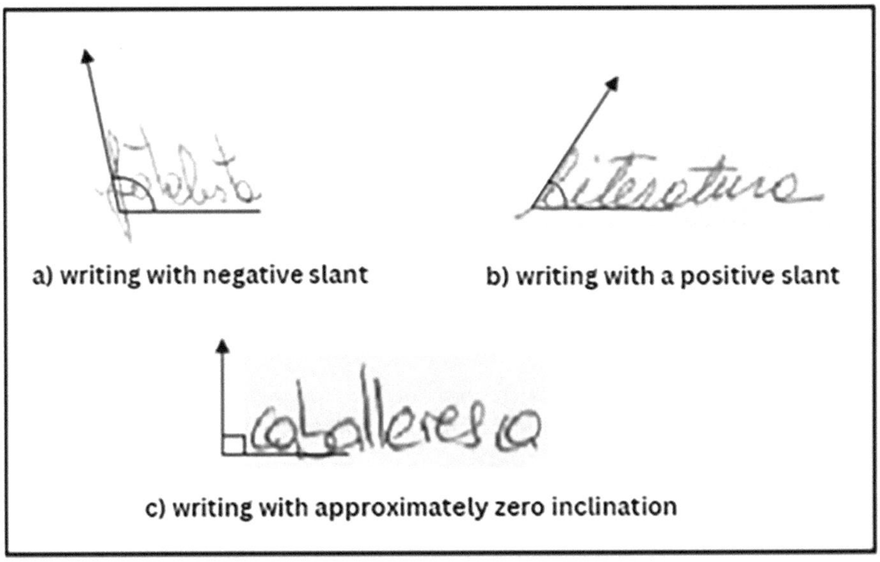

Figure 8.
Examples of inclination types.

3.3.4 Colligation

Colligation, also known as cohesion, refers to the degree to which the writing appears connected. It should be noted that colligation is not synonymous with continuity, although there may be cases where the two concepts overlap. In offline processing, it becomes challenging to determine whether a person lifts the writing instrument, as a stroke may lack separation and appear to have been written in parts.

To assess the type of cohesion in a writing sample, the predominant percentage needs to be determined. Based on this percentage, we can classify the writing using the following guidelines:

Linked writing: Letters and parts of letters are interconnected without any breaks in between. The movements exhibit cohesion, remaining consistent and uninterrupted. However, cohesion or linking may be momentarily interrupted for essential strokes such as dots, accents, capital letters, or the horizontal stroke of the letter "T."

Unlinked or juxtaposed writing: Words consist of unlinked letters, where there is no contact between letters (although they may touch). Words are independent, lacking cohesion. Grouped writing: Words are formed by groups of two, three, four, or more letters, depending on the word's length.

Fragmented or disjointed writing: This term describes writing where letters are composed of two or more separate (fragmented) strokes. Capital letters and the letter "m" can give the impression of disjointed movements. By assessing the cohesion in writing samples, handwriting experts can gain insights into an individual's writing style, providing valuable information for analysis and comparison purposes (**Figure 9**).

Algorithmically, when characterizing the cohesion parameter, each image contains groups of active pixels (foreground) or connected components with an 8-neighbor connectivity. Due to the offline nature of the analysis, where temporal sequencing is

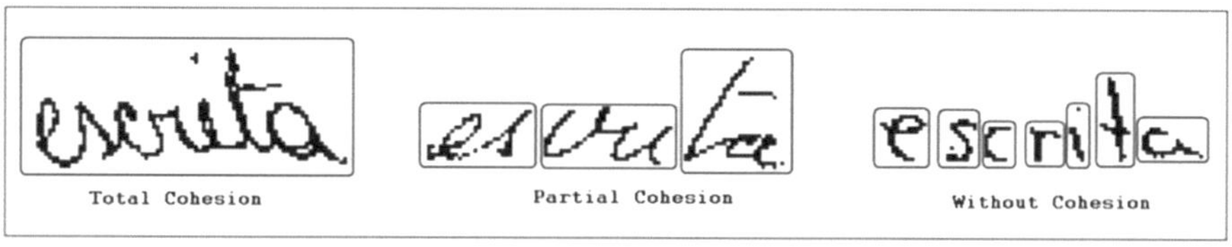

Figure 9.
Representation of three types of colligations.

not available to indicate the writing process, we can only differentiate whether individuals write words in a linked or separate manner. We can identify cases of partially linked or completely unlinked writing. However, we are unable to discern situations where the writer lifts the tool between each letter but maintains complete linkage.

This expression serves as a quantitative measure to evaluate the cohesion level within a writing sample.

$$f_4 = 1 - \frac{\text{Number of separations}}{\text{Number of connected components}} \tag{6}$$

3.4 Experimental methodology

The methodology employed to accomplish the objectives consists of four key steps, which are outlined below.

Firstly, a comprehensive set of characteristics commonly employed in forensic handwriting analysis was extracted. It should be noted that the coding of these parameters did not always align with the practices of this discipline. Specifically, qualitative classifications of certain characteristics were quantitatively estimated, aiming to capture the essence of the established categories. Additionally, word contour coding was performed using two techniques: the application of Fischer's Kernel to the Markovian representation of the contour and Fourier descriptors.

Subsequently, the graphometry parameters underwent an analysis of variance to determine their level of consistency. It is crucial to highlight that the database utilized in this stage was meticulously constructed under controlled conditions. This approach ensures the accurate retrieval of the inherent nature of the characteristics, mitigating any potential interference in the analysis caused by noise generated from the context. By conducting the analysis under controlled conditions, the reliability and validity of the results were maximized, providing a robust foundation for further investigations.

In the third step, the effectiveness of the graphometric parameters and contour encodings was evaluated. Specifically, the results obtained from various classifiers in the task of writer recognition were thoroughly analyzed. This analysis aimed to assess the discriminatory power and accuracy of the selected parameters and encoding techniques in distinguishing between different writers. By examining the performance of different classifiers, including but not limited to neural networks, Support Vector Machines, and decision trees, valuable insights were gained regarding the effectiveness of the graphometry features in achieving accurate and reliable writer identification. The findings from this step contributed to the refinement and optimization of the proposed methodology, enhancing its overall efficacy in individual identification based on handwriting analysis.

In the final stage of this methodology, a selection and validation process were conducted to identify the most persistent and discriminant graphometry parameters.

The validation was performed by applying these parameters in the task of writer recognition. In the training phase, a model for each writer was established using recent writing samples, while in the test phase, writings that were at least 10 years old and were not included in the training phase were utilized. This approach aimed to assess the reliability and longevity of the selected parameters by evaluating their effectiveness in accurately identifying writers based on handwriting samples that exhibited significant temporal variations. The validation process ensured that the chosen parameters were robust and capable of maintaining their discriminatory power over extended periods of time, strengthening the overall validity and practicality of the proposed methodology.

4. Results

This section presents a detailed analysis of the experimental results obtained from the application of various classification algorithms, namely K-Nearest Neighbors (KNN), Multilayer Perceptron (MLP), Neural Network (NN), and Support Vector Machines (SVM).

Table 2 presents the classification accuracies obtained for different parameter combinations related to slant (Slant HJ, Slant BS), length (Length HJ), concentration (Concentration), roundness (Roundness), final features (End features, Final traits), colligation (Colligation), and pressure (Pressure). The results demonstrate the influence of these parameters on the performance of the classification models.

For the parameter combination of Slant HJ, Direction; Length HJ, Concentration; Roundness, Final features; Colligation; Pressure, the classification accuracies were 95.70 ± 1.94% for KNN, 95.76 ± 1.48% for NN, and 89.1 ± 2.60% for SVM. This indicates that all three models achieved relatively high accuracy in classifying the samples based on these combined parameters.

Similarly, the parameter combination of Slant BS, Direction; Length HJ, Concentration; Roundness, Final features; Colligation; Pressure yielded accuracies of 95.86 ± 1.69% for KNN, 95.46 ± 1.57% for NN, and 88.98 ± 2.16% for SVM. These results suggest that the models performed consistently across different parameter combinations, with KNN and NN achieving slightly higher accuracies compared to SVM.

	KNN	MLP	SVM
Slant HJ, Direction; Length HJ, Concentration; Roundness, Final features; Colligation; Pressure	95.70 ± 1.94	95.76 ± 1.48	89.1 ± 2.60
Slant BS, Direction; Length HJ, Concentration; Roundness, Final features; Colligation; Pressure	95.86 ± 1.69	95.46 ± 1.57	88.98 ± 2.16
Slant HJ, Direction; Length HJ, Concentration; Roundness, Final features; Colligation; Pressure	95.92 ± 1.04	96.18 ± 1.24	90.24 ± 2.08
Slant HJ, Slant BS; Length HJ, Concentration; Roundness, Final features; Colligation; pressure	96.36 ± 0.92	96.32 ± 1.61	90.08 ± 2.15

Table 2.
Hit rate by including two parameters in some of the graphical elements: (Slope; Dimension; Richness; Link; Pressure). 50% training.

Furthermore, the combination of Slant HJ, Direction; Length HJ, Concentration; Roundness, Final features; Colligation; Pressure demonstrated improved performance, with accuracies of 95.92 ± 1.04% for KNN, 96.18 ± 1.24% for NN, and 90.24 ± 2.08% for SVM. These findings indicate the effectiveness of the selected parameters in accurately classifying the handwriting samples.

Moreover, when additional parameters such as Slant BS and End features were considered in the parameter combination of Slant HJ, Slant BS; Length HJ, Concentration; Roundness, Final features; Colligation; Pressure, the classification accuracies further increased. The models achieved accuracies of 96.36 ± 0.92% for KNN, 96.32 ± 1.61% for NN, and 90.08 ± 2.15% for SVM. These results highlight the importance of including a comprehensive set of parameters in achieving higher classification accuracies.

To delve deeper into the analysis, **Table 3** presents the influence of specific combinations of graphometry parameters on the classification performance. For instance, when considering slant, length, roundness, colligation, and pressure, KNN achieved an accuracy of 91.84 ± 1.64%, while NN achieved an accuracy of 91.40 ± 1.75%, and SVM yielded 77.62 ± 3.33%. Similar trends were observed for other parameter combinations.

Based on the provided tables, the model that consistently demonstrates the highest classification accuracy across different parameter combinations is the K-Nearest Neighbors (KNN) algorithm. In both tables, KNN consistently achieves the highest or one of the highest accuracies among the three classifiers (KNN, MLP, and SVM).

These results highlight the effectiveness of the proposed classification algorithms in accurately identifying and distinguishing various handwriting characteristics. The outstanding accuracies obtained when utilizing all parameters in a text-independent scenario emphasize the importance of incorporating a comprehensive set of graphometry features. Furthermore, the varying performance across different parameter combinations suggests that certain combinations possess stronger discriminative capabilities.

The findings from this study provide valuable insights into the potential of graphometry and classification algorithms for achieving precise and reliable writer recognition. However, further investigation and optimization are required to enhance the overall performance and generalizability of the proposed approach. Future research could focus on exploring alternative combinations of graphometry parameters, investigating advanced feature extraction techniques, and refining the training process of the classification algorithms.

	KNN	NN	SVM
Slant HJ; Length HJ; Roundness; Colligation; Pressure.	91.84 ± 1.64	91.40 ± 1.75	77.62 ± 3.33
Slant BS; Length HJ; Roundness; Colligation; Pressure.	91.90 ± 1.61	91.90 ± 2.09	78.16 ± 2.74
Direction; Length HJ; Roundness; Colligation; Pressure.	83.00 ± 1.62	77.40 ± 3.51	66.18 ± 3.10
Slant HJ; Length HJ; End features; Colligation; pressure	92.36 ± 1.57	91.00 ± 2.71	82.96 ± 2.35
Slant HJ; Calligraphic box; Final traits; Colligation; Pressure.	90.30 ± 1.94	88.20 ± 3.23	80.02 ± 2.70
Slant HJ; Concentration; Final features; Colligation; Pressure.	91.00 ± 1.69	88.50 ± 3.03	80.40 ± 2.92
Slant HJ; Length HJ; grapheme "a"; colligation; pressure.	90.18 ± 1.31	91.50 1.73	86.08 ± 1.59

Table 3.
Hit rate by including one parameter in some of the graphical elements: (Slope; Dimension; Richness; Link; Pressure). 50% training.

References	Dataset script	Technique used	Accuracy (%)
Purohit et al. [36]	IAM English language dataset	CNN	92.7
Hagstrom et al. [37]	4920 written DoBs (own dataset)	ResNet50	94
Nabi et al. [38]	Urdu (own dataset)	DeepNet-WI (VGG-16)	98.71
Javidi and Jampour [39]	IAM, CERUG, FIREMAKER, and CVL	Res-Net	88.95
Proposed technique	350 Spanish words	KNN	96.36

Table 4.
Comparison of the proposal vs. the state of the art of the proposal.

These results contribute significantly to the existing body of knowledge in the field of writer recognition and open possibilities for practical applications in forensic studies, document analysis, and other relevant domains. The promising outcomes encourage further exploration of graphometry analysis techniques to fully unlock their potential in the realm of handwriting expertise.

4.1 Comparison with similar works

Among the referenced studies, Purohit et al. [36] utilized the IAM English language dataset and employed a CNN technique, achieving an accuracy of 92.7%. Hagstrom et al. [37] conducted their research using a dataset consisting of 4920 written dates of birth (DoBs) and applied the ResNet50 technique, resulting in an accuracy of 94%.

Nabi et al. [38] worked with an Urdu dataset they developed themselves and utilized the DeepNet-WI (VGG-16) technique, achieving an impressive accuracy of 98.71%.

Javidi and Jampour [39] performed their analysis using multiple datasets including IAM, CERUG, FIREMAKER, and CVL. They employed the Res-Net technique and obtained an accuracy of 88.95%.

In comparison to these studies (see **Table 4**), our proposed technique involved a dataset of 350 Spanish words, and we employed the KNN technique. Our results yielded a high accuracy of 96.36%. It is noteworthy that our technique outperformed the accuracies achieved by [36, 37, 39], while being slightly below the exceptional accuracy achieved by [38] in their Urdu dataset.

Overall, these findings indicate the effectiveness of our proposed technique in handwriting recognition, particularly for Spanish words, and highlight its competitive performance when compared to previous studies using different datasets and techniques.

5. Conclusions

In conclusion, this academic article underscores the crucial role of handwriting as a biometric-behavioral trait in person recognition, particularly within the realm of forensic studies focused on document authorship determination. Through an extensive exploration of various feature extraction approaches and the utilization of diverse classifiers, the research presented in this article has significantly advanced the understanding and application of handwriting analysis for accurate identification outcomes.

The study delved into a wide array of feature extraction techniques, encompassing contour-based features, interpolation-based curvature information, wavelet transforms, and Hermite coefficients. These approaches showcased the versatility and effectiveness of different methods in capturing distinctive handwriting characteristics essential for reliable identification.

In addition, the investigation highlighted the significance of employing different classifiers, including artificial neural networks, k-nearest neighbors, Support Vector Machines, Hidden Markov Models, and Gaussian mixture models. By leveraging these classifiers, the study demonstrated the robustness and adaptability of various classification techniques in accurately discerning individual handwriting patterns.

Moreover, the research shed light on the vital role of morphometric and structural features in forensic document analysis, providing insights into their relevance for both establishing identity and unraveling psychological profiles. These features, categorized as quantitatively evaluated parameters and qualitatively analyzed characteristics, present both challenges and opportunities for automated processing and serve as avenues for future research and development.

Throughout the experimental phase, the study successfully grouped parameters into distinct grapheme elements. Notably, the combination of parameters within the inclination, dimension, and shape richness elements yielded remarkable hit rates surpassing 50%. While individual grapheme elements such as link and pressure exhibited consistent but relatively lower hit rates, they nevertheless contributed to the overall understanding of handwriting characteristics.

To summarize, this comprehensive investigation significantly advances the field of handwriting-based biometric identification. By exploring various feature extraction approaches, employing diverse classifiers, and emphasizing the importance of graphometry features, the study paves the way for the development of robust and sophisticated systems capable of accurately recognizing individuals based on their handwriting. The findings hold substantial value for forensic document analysis, offering insights and implications for applications requiring precise writer identification in a variety of domains.

In conclusion, this research lays a solid foundation for further advancements in the field, propelling the development of cutting-edge systems that leverage the intricacies of handwriting for accurate identification. The knowledge gained from this study holds great promise for enhancing forensic investigations, bolstering security systems, and enabling various other domains to benefit from the reliable and precise identification of individuals through their unique handwriting patterns.

Further results obtained by merging deep learning systems together with our machine learning system for manuscript identification will be included as proposed future lines of research.

Author details

Jose Luis Vásquez-Vasquez[1] and Carlos M. Travieso-González[2]*

1 Universidad de Costa Rica, Turrialba, Costa Rica

2 Department of Signals and Communication, Institute for Technological Development and Innovation in Communications, University of Las Palmas, Gran Canaria, Spain

*Address all correspondence to: carlos.travieso@ulpgc.es

References

[1] Srihari SN, Cha SH, Arora H, Lee S. Individuality of handwriting: A validation study. In: Proceedings of the International Conference on Document Analysis and Recognition, ICDAR. 2001-January, 2001. pp. 106-109. DOI: 10.1109/ICDAR.2001.953764

[2] Srihari SN, Xu Z, Hanson L. Development of handwriting individuality: An information-theoretic study. In: Proceedings of International Conference on Frontiers in Handwriting Recognition, ICFHR. Vol. 2014-December. 2014. pp. 601-606. DOI: 10.1109/ICFHR.2014.106

[3] Schomaker L. Advances in writer identification and verification. In: Proceedings of the International Conference on Document Analysis and Recognition, ICDAR. Vol. 2. 2007. pp. 1268-1273. DOI: 10.1109/ICDAR.2007.4377119

[4] Siddiqi I, Vincent N. A set of chain code based features for writer recognition. Proceedings of the International Conference on Document Analysis and Recognition, ICDAR. 2009:981-985. DOI: 10.1109/ICDAR.2009.136

[5] Chanda S, Franke K, Pal U. Text independent writer identification for Oriya script. In: Proceedings—10th IAPR International Workshop on Document Analysis Systems, DAS. Vol. 2012. 2012. pp. 369-373. DOI: 10.1109/DAS.2012.86

[6] Hiremath PS, Shivashankar S, Pujari JD, Mouneswara V. Script identification in a handwritten document image using texture features. In: 2010 IEEE 2nd International Advance Computing Conference, IACC. Vol. 2010. 2010. pp. 110-114. DOI: 10.1109/IADCC.2010.5423028

[7] Imdad A, Bres S, Eglin V, Emptoz H, Rivero-Moreno C. Writer identification using steered hermite features and SVM. Proceedings of the International Conference on Document Analysis and Recognition, ICDAR. 2007;2:839-843. DOI: 10.1109/ICDAR.2007.4377033

[8] Anton C, Stirbu C, Badea RV. Automatic hand writer identification using the feed forward neural networks. In: World Congress on Internet Security, WorldCIS-2011. 2011. pp. 290-293. DOI: 10.1109/WORLDCIS17046.2011.5749871

[9] Chaturvedi S, Titre RN, Sondhiya N. Review of handwritten pattern recognition of digits and special characters using feed forward neural network and izhikevich neural model. In: Proceedings—International Conference on Electronic Systems, Signal Processing, and Computing Technologies, ICESC. Vol. 2014. 2014. pp. 425-428. DOI: 10.1109/ICESC.2014.83

[10] Marti UV, Messerli R, Bunke H. Writer identification using text line based features. In: Proceedings of the International Conference on Document Analysis and Recognition, ICDAR. Vol. 2001-January. 2001. pp. 101-105. DOI: 10.1109/ICDAR.2001.953763

[11] Ibrahim AS, Youssef AE, Abbott AL. Global vs. local features for gender identification using Arabic and English handwriting. In: 2014 IEEE International Symposium on Signal Processing and Information Technology, ISSPIT 2014. 2015. DOI: 10.1109/ISSPIT.2014.7300580

[12] Djeddi C, Meslati LS, Siddiqi I, Ennaji A, El Abed H, Gattal A. Evaluation of texture features for offline

Arabic writer identification. In: Proceedings—11th IAPR International Workshop on Document Analysis Systems, DAS. Vol. 2014. 2014. pp. 106-110. DOI: 10.1109/DAS.2014.76

[13] Schlapbach A, Bunke H. Off-line handwriting identification using HMM based recognizers. In: Proceedings—International Conference on Pattern Recognition. Vol. 2. 2004. pp. 654-655. DOI: 10.1109/ICPR.2004.1334343

[14] Christlein V, Bernecker D, Hönig F, Maier A, Angelopoulou E. Writer identification using GMM supervectors and exemplar-SVMs. Pattern Recognition. 2017;**63**:258-267. DOI: 10.1016/J.PATCOG.2016.10.005

[15] Slimane F, Märgner V. A new text-independent GMM writer identification system applied to Arabic handwriting. In: Proceedings of International Conference on Frontiers in Handwriting Recognition, ICFHR. Vol. 2014-December. 2014. pp. 708-713. DOI: 10.1109/ICFHR.2014.124

[16] Rafiee A, Motavalli H. Off-line writer recognition for farsi text. In: Proceedings–2007 6th Mexican International Conference on Artificial Intelligence, Special Session, MICAI. Vol. 2007. 2007. pp. 193-197. DOI: 10.1109/MICAI.2007.37

[17] Pervouchine V, Leedham G. Extraction and analysis of forensic document examiner features used for writer identification. Pattern Recognition. 2007;**40**(3):1004-1013. DOI: 10.1016/J.PATCOG.2006.08.008

[18] Pastor Gadea M. Aportaciones al reconocimiento automático de texto manuscrito. 2007. Available from: https://dialnet.unirioja.es/servlet/tesis?codigo=17935&info=resumen&idioma=SPA [Accessed: June 12, 2023 (Online)]

[19] Jain R, Doermann D. Combining local features for offline writer identification. In: Proceedings of International Conference on Frontiers in Handwriting Recognition, ICFHR. Vol. 2014-December. 2014. pp. 583-588. DOI: 10.1109/ICFHR.2014.103

[20] Angadi SA, Angadi SA, Angadi SH. Structural features for recognition of hand written Kannada character based on SVM biometrics view project research_hatture view project. Article in International Journal of Computer Science Engineering and Information Technology. 2015;**5**(2). DOI: 10.5121/ijcseit.2015.5203

[21] Abdulrahman SA, Alhayani B. A comprehensive survey on the biometric systems based on physiological and behavioural characteristics. Materials Today: Proceedings. 2023;**80**:2642-2646. DOI: 10.1016/J.MATPR.2021.07.005

[22] Handbook of Biometrics—Google Libros. Available from: https://books.google.es/books?hl=es&lr=&id=WfCowMOvpioC&oi=fnd&pg=PA1&dq=A.K.+Jain%3B+P.+Flynn%3B+A.A.+Ross%3B+%E2%80%9CHandbook+of+biometrics%E2%80%9C+,+Springer,+ISBN-13:+978-0-387-71040-2,+USA,+2007.&ots=xrXI5Tx5Gf&sig=QO7bkHWtNuz95wCzmVtLUj8U2is&redir_esc=y#v=onepage&q&f=false [Accessed: June 12, 2023]

[23] Bozinovic RM, Srihari SN. Off-line cursive script word recognition. IEEE Transactions on Pattern Analysis and Machine Intelligence. 1989;**11**(1):68-83. DOI: 10.1109/34.23114

[24] de Sousa Neto AF, Bezerra BLD, Toselli AH, Lima EB. A robust handwritten recognition system for learning on different data restriction scenarios. Pattern Recognition Letters. 2022;**159**:232-238. DOI: 10.1016/J.PATREC.2022.04.009

[25] Raj MAR, Abirami S, Shyni SM. Tamil handwritten character recognition system using statistical algorithmic approaches. Computer Speech & Language. 2023;**78**:101448. DOI: 10.1016/J.CSL.2022.101448

[26] Zhang G, Wang W, Zhang C, Zhao P, Zhang M. HUTNet: An Efficient Convolutional Neural Network for Handwritten Uchen Tibetan Character Recognition. 2023. DOI: 10.1089/BIG.2021.0333. Available from: https://home.liebertpub.com/big

[27] Contreras Gedler JA. Cognitive cryptography using behavioral features from linguistic-biometric data. Cryptology ePrint Archive. 2023

[28] Li X, Cervantes J, Yu W. A novel SVM classification method for large data sets. In: Proceedings—2010 IEEE International Conference on Granular Computing, GrC. Vol. 2010. 2010. pp. 297-302. DOI: 10.1109/GRC.2010.46

[29] Cervantes J, Li X, Yu W, Bejarano J. Multi-class support vector machines for large data sets via minimum enclosing ball clustering. In: 2007 4th International Conference on Electrical and Electronics Engineering, ICEEE 2007. 2007. pp. 146-149. DOI: 10.1109/ICEEE.2007.4344994

[30] Banerjee S. Boosting inductive transfer for text classification using Wikipedia. In: Proceedings—6th International Conference on Machine Learning and Applications, ICMLA. Vol. 2007. 2007. pp. 148-153. DOI: 10.1109/ICMLA.2007.25

[31] Cervantes J, Li X, Yu W. Support vector classification for large data sets by reducing training data with change of classes. In: Proceedings of IEEE International Conference on Systems, Man and Cybernetics. 2008. pp. 2609-2614. DOI: 10.1109/ICSMC.2008.4811689

[32] Support Vector Machines—Ingo Steinwart, Andreas Christmann - Google Libros. Available from: https://books.google.es/books?hl=es&lr=&id=HUnqnrpYt4IC&oi=fnd&pg=PA1&dq=Steinwart+and+Christmann,+2008&ots=gakJEu1sUa&sig=JC39H7zWWeQdIT53Mi5YlkxL_F0&redir_esc=y#v=onepage&q=Steinwart%20and%20Christmann%2C%202008&f=false [Accessed: June 22, 2023]

[33] Gatos B, Papamarkos N, Chamzas C. Skew detection and text line position determination in digitized documents. Pattern Recognition. 1997;**30**(9):1505-1519. DOI: 10.1016/S0031-3203(96)00157-4

[34] Kirli Ö, Gülmezoğlu MB. Automatic writer identification from text line images. International Journal on Document Analysis and Recognition. 2012;**15**(2):85-99. DOI: 10.1007/S10032-011-0161-9/METRICS

[35] Otsu N. Threshold selection method from gray-level histograms. IEEE Transactions on Systems, Man, and Cybernetics. 1979;**SMC-9**(1):62-66. DOI: 10.1109/TSMC.1979.4310076

[36] Purohit N, Panwar S. Dual-pathway deep CNN for offline writer identification. Lecture Notes in Networks and Systems. 2022;**249**:119-127. DOI: 10.1007/978-3-030-85365-5_12/COVER

[37] Hagstrom AL, Stanikzai R, Bigun J, Alonso-Fernandez F. Writer Recognition Using Off-line Handwritten Single Block Characters. In: 2022 International Workshop on Biometrics and Forensics (IWBF). 2022. DOI: 10.1109/IWBF55382.2022.9794466

[38] Nabi ST, Kumar M, Singh P. DeepNet-WI: A deep-net model for offline Urdu writer identification. Evolving Systems. 2023;**1**:1-11. DOI: 10.1007/S12530-023-09504-/TABLES/3

[39] Javidi M, Jampour M. A deep learning framework for text-independent writer identification. Engineering Applications of Artificial Intelligence. 2020;**95**:103912. DOI: 10.1016/J.ENGAPPAI.2020.103912

Chapter 9

Biometric-Based Optical Systems for Security and Authentication

Gaurav Verma, Wenqi He and Xiang Peng

Abstract

In a digital world, biometric authentication is becoming more and more popular for reliable automatic recognition of people, which is widely being deployed in optical information security-related systems. The adoption of biometrics into optical security-based applications and fields has been adding excellent security due to their distinctive attribute that gains from optics. In this chapter, we present an optical nonlinear cryptosystem for image encryption using biometric keys generated from fingerprint hologram for security and authentication. In order to generate biometric keys, we implemented an optoelectronics experiment setup using digital holography for capturing the fingerprint hologram, storing, and then numerically reconstructing it. The reconstructed features of the fingerprint object offer very appealing attributes from the perspective of data encryption such as uniqueness, randomness, and discriminability. Fingerprint biometric features are kept inside interference patterns optically, which are also protected with experimental parameters. If both pieces of information are provided to be known to the person at the decryption stage, as a result, it keeps maintaining user specificity in order to access system information. Furthermore, we exploit the utility of the biometric key in designing an optical cryptosystem for encrypting the information which offers a solution to the distribution of keys with heightened security.

Keywords: fingerprint biometric, optical encryption, security, authentication, digital holography

1. Introduction

Biometrics refers to a unique, measurable, biological trait or attribute of a human being that is used to validate the identity of a person [1, 2]. The use of biometrics in a system relies on automated recognition methods based on a person's physiological or behavioral features, whose functionality works on the conforming exact identity of an individual compared to traditional authentication methods such as passwords, tokens, and PINs (Personal Identification Numbers) [3–5]. A number of applications are implemented to automate authentication methods by the use of biometric traits for access control, commercial, phone, government, and forensic [1–6]. In general, the biometric trait comprises physiological or behavioral features [5]. The physiological traits, which use fingerprints, retina, iris, facial images, and hand geometry, are physical characteristics computed at a specific point in time, while behavioral biometric traits commonly list in particular, signature, gait, voice recordings, and keystroke rhythms, make attention to the mode some action is accomplished by every individual [1, 4].

Due to recent technological advances in real-world applications, an automated identification system has been implemented in many security-related systems using biometric traits for reliable and trusted authentication [1–7]. Moreover, different kinds of threats, challenges, and privacy issues are growing concerns in today's modern world, and biometric technology is used to ensure secure and safe circumstances [4–7]. The development of optics-based biometric systems has brought tremendous growth in data security and authentication in recent times [6, 7]. More excitingly, most of the current optical encryption systems are now being incorporated with the biometric features of a person [7]. These scenarios, however, increase the security level for information protection through person identification against unauthorized access [4].

Optical systems are extensively involved in the field of information security due to high speed, parallel processing, and exploitation of multidimensional data such as wavelength, frequency, and polarization [7–66]. Optics-related cryptosystems, such as optical, compression, encryption, photon counting, and authentication, are extensively developed to secure sensitive data or images during transmission and reception through the digital medium [6, 7]. In the field of optics, image encryption was brought into existence since the introduction of the double random phase encoding (DRPE) scheme. The DRPE method converts input information into white stationary noise by the involvement of two random phase masks (RPMs) at the input and the Fourier plane, respectively [8]. Due to the linear and symmetric nature of the DRPE system, these RPMs for encryption and decryption processes are similar to security keys [8–13]. In order to take advantage of optical encryption, Qin *et al.* reported the optical cryptosystem for image encryption based on phase-truncated Fourier transforms (PTFTs) [14, 15]. The PTFT operation is used to truncate the Fourier spectrum of the image into the phase and the amplitude distributions. The PTFT encryption scheme uses two RPMs for encryption, while two phase-only masks are obtained as the decryption keys for the decryption process. The main advantages of PTFT over the DRPE are that keys for encryption and decryption processes are distributed due to nonlinear operation. These RPMs act as the main security component. From the cryptanalysis point of view, it is noticed that the RPM-based encryption system suffers from the problem of key management, distribution, and authentication and is found to be insecure against various types of attacks [12, 13, 16–19]. Furthermore, by adding security to optical encryption using unique features of human beings, optical security using biometrics facilitates a secure and reliable way for information processing. Many types of optical systems using biometrics are implemented [24–40], which offer a wide range of features such as keys management, higher security, and user authentication.

From the recent research in the optical encryption domain, it is observed that optical cryptosystems are perceived to be unsuited to implement traditional cryptography because of being unable to address the key distribution issue in terms of public keys and private keys [33–53]. A nonlinear encryption system in the Fresnel domain using the optical phase-retrieval algorithm is developed as reported in [39], which fulfills the criteria of asymmetric cryptography agreement. The phase-retrieval algorithm-based method has also been studied on the development of nonlinear cryptosystems that show progress in the generation of public and private keys in the encryption system, which is demanded with authentication. Zhao *et al*. presented an optical nonlinear cryptosystem using a fingerprint combined with a phase-retrieval algorithm and public key cryptography [40]. In this scheme, the fingerprint features of a person are associated with encryption and decryption operations that help to decrypt the information in the authenticated way at the receiver and also solve the issue of the public-private keys [40–42]. One of the approaches to the implementation of optical information authentication systems is carried out by combining a median-filtering-based phase-retrieval algorithm [51]. Moreover, optical image encryption techniques using digital holography are applied for authentication and security [32–34, 54–65]. This method includes an optical process for recording a hologram by the involvement of a charge-coupled device (CCD), and the captured hologram is known as a digital hologram, which is further numerically reconstructed in a computer [32–34, 57]. The reconstructed information provides additional information in terms of the amplitude, and the phase of the object. Thus, the reconstructed features are explored in the design of optical information processing systems for security and authentication.

In this chapter, we describe our proposed optical nonlinear cryptosystem for image encryption using biometric keys based on an optical phase-retrieval algorithm and phase-truncated Fourier transform for security and authentication, which offer the solution of key distribution with improved security. In this direction, we implement the optoelectronics experiment based on digital holography for recording the fingerprint hologram, which is digitally reconstructed to obtain keys information in terms of the amplitude and the phase. The merit is that the digital recording and reconstruction process of the fingerprint hologram using holography makes it possible for transmission and reception over a communication medium. In addition, the fingerprint hologram is protected by a reconstruction parameter which also enables verification and authentication approach in the proposed encryption/decryption processes. First, we introduce the idea of the optoelectronic experimental process for recording the fingerprint hologram, and its numerical reconstruction. Next, we analyze the features of the reconstructed fingerprint image by performing the statistical test that makes its usage as an encryption key for the image [32]. Furthermore, we explore the utility of the biometric key for optical cryptosystem for image encryption based on the phase-retrieval algorithm and the phase-truncated Fourier transforms (PTFT) scheme. The system uses keys for encrypting the information using the public keys or encryption keys that can only be truly recovered using the private keys or decryption keys, while the involvement of biometric keys maintains the authenticity of the user throughout the process. Our work is the first attempt to develop an optical cryptosystem using the phase-retrieval algorithm and PTFT combined image encryption system utilizing biometric keys from fingerprint hologram. Finally, we demonstrate the security performance and robustness of our cryptosystem. This chapter is structured as follows: Section 2 introduces the optoelectronics setup using digital holography for biometric keys generation and analyzes the keys features demonstrating its utility for image encryption. Section 3 introduces the cryptography perspective using biometrics. Section 4 presents an implementation of optical encryption process. Section 5 investigates experimental results to present enhanced security with user authentication and management of keys. Finally, the conclusion presents the significant contributions, discusses the challenges of the work, and suggests future research directions.

2. Optoelectronic experimental setup for biometric keys

In this section, the method of fingerprint hologram recording has been presented using digital holography in order to capture both phase and amplitude distributions [54–65]. In order to perform fingerprint imaging, an optoelectronics experiment using a digital holographic technique is implemented, as shown in **Figure 1**.

2.1 Fingerprint database

From the biometric perspective, fingerprints contain essential patterns like arches, ridges, and whorls on the surface of a finger which are unique to the person. In our study, we use samples of the fingerprint from the fingerprint verification competition (FVC) database, as reported in [66]. The 'FVC' term signifies a fingerprint verification competition. This database has eight sets of fingerprint impressions of 100 users which are captured and collected using different sensor-based technologies. Features of the fingerprint images from the dataset have been extracted with pixels 300 × 480 and a resolution of 512 dots per inch (dpi). From an imaging perspective, the fingerprint is employed in a digital holographic-based setup as shown in **Figure 1** and its detailed process is explained in Section 2.2.

2.2 Recording process of the biometric hologram

In order to capture fingerprint image hologram using the experimental setup as shown in **Figure 1**, the optical beam emerging from the He-Ne laser is initially collimated through spatial filtering (SF) operation and then separated into the object arm and the reference arm with the use of a beam splitter (BS_1), which is directed with the help of mirrors (Ms) and finally combined at the BS_2. The light that passes from the fingerprint object is known as the object beam, while another beam O(x,y) that comes from the reference arm is denoted as the reference beam U(x,y). Several techniques for the acquisition of fingerprints are widely investigated by researchers in security and optical imaging-related applications. In our optical configuration, we use fingerprint images of a person from a public biometric dataset as reported in [66]. In order to perform transmission imaging, fingerprint features are displayed on the transparent sheet of size '1 cm × 1 cm' which shows high contrast features for

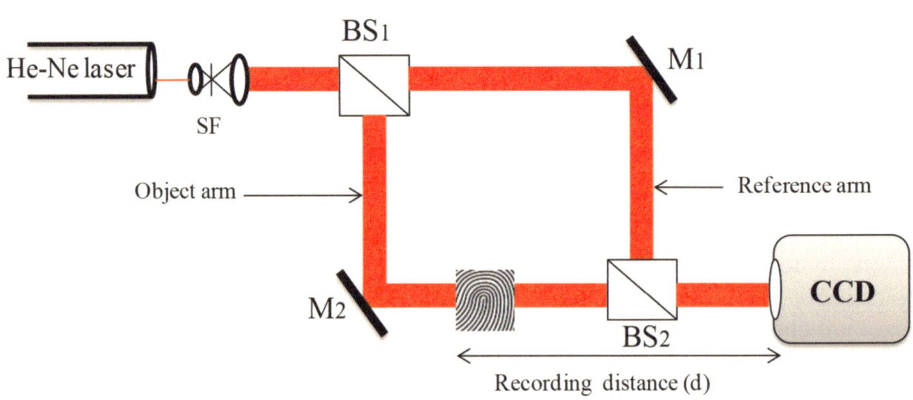

Figure 1.
Optoelectronic experimental setup: BSs: beam splitters, Ms: mirrors, SF: spatial filtering.

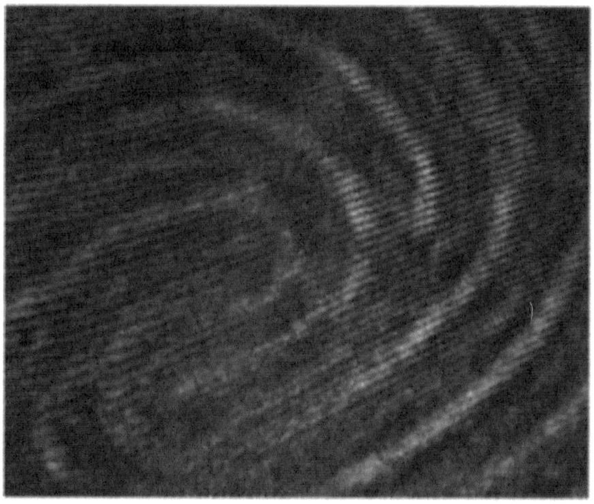

Figure 2.
Recorded fingerprint hologram.

recording as per the detailed procedure reported in our previous work [32, 33, 57]. In the object arm, the fingerprint images of size '1 cm × 1 cm' are employed at a distance'*d*' from the charge-coupled device (CCD) device [32, 33, 57]. As a result, this optical configuration makes it practically more suitable for optical imaging as well as information processing using digital holographic techniques in comparison to conventional fingerprint acquisition methods as described in [57]. In the process of recording, when light rays from a He-Ne laser strike the fingerprint object it causes diffraction, scattering, and absorption. This phenomenon carries signifying information about the object's amplitude and shape, which further interfered with the reference beam that is recorded by the CCD camera. The recorded interference pattern contains the complete information on the fingerprint object shown in **Figure 2**, which is stored in the computer and coined as a digital hologram. This process can be mathematically expressed as:

$$\begin{aligned} H(x,y) &= |O(x,y) + U(x,y)|^2 \\ &= O(x,y)O^*(x,y) + U(x,y)U^*(x,y) \\ &\quad + O(x,y)U^*(x,y) + U(x,y)O^*(x,y) \end{aligned} \tag{1}$$

As given in Eq. (1), the '$H(x,y)$' is termed as the digital hologram and '*' shows the complex conjugate operation. The recording parameters for the fingerprint hologram are given as wavelength λ = 632.8 *nm*, the distance d = 0.29 *m*, and pixel sizes 4.65 μm × 4.65 μm of the CCD (Lumenera's Infinity2, 1360 × 1024 pixels).

2.3 Reconstruction process of the fingerprint hologram

In order to reconstruct the fingerprint features from the recorded hologram as shown in **Figure 2**, the reconstruction processes using the Fresnel-Kirchhoff integral are employed numerically, which makes it free from the zero order term in separating the real and the virtual images [32, 33, 54–57] as:

$$D(v,u) = \frac{i}{\lambda}\int\int_{-\infty}^{\infty} H(x,y)\, E_R(x,y)\, \frac{\exp\left(-i\,\frac{2\pi}{\lambda}\rho'\right)}{\rho'}\, dx\, dy \tag{2}$$

$$\rho' = \sqrt{(x-v')^2 + (y-u')^2 + d^2} \tag{3}$$

where $D(v, u)$ shows the reconstructed object, E_R refers to the plane wave, and ρ' represents the distance between the recording plane (x, y) and the reconstruction plane (v', u'), respectively. It can be noticed that the process of reconstruction is completely carried out digitally. From the reconstructed object as demonstrated in **Figure 3**, the real fingerprint image of size 146 × 146 pixels is extracted in our analysis. The reconstructed fingerprint object results in a complex field which is further separated in terms of the intensity and phase distribution:

$$D(v,u) = A(v,u).exp\,(i\phi(v,u)) \tag{4}$$

$$A(v,u) = \mathrm{Re}\left[D(v,u)^2\right] + \mathrm{Im}\left[D(v,u)^2\right] \tag{5}$$

$$\varnothing(v,u) = arctan\,\frac{\mathrm{Im}[D(v,u)]}{\mathrm{Re}[D(v,u)]} \tag{6}$$

The reconstructed object clearly reproduces the fingerprint pattern as shown in **Figure 3a** while the phase obtained from the same hologram is represented in **Figure 3b** which shows significant information on the fingerprint pattern as well as some variation of thickness over the field of view. Therefore, both pieces of information clearly exhibit the significant features of the fingerprint biometric.

Furthermore, it is evident that the obtained phase is utilized to construct a phase mask (PM) with phases uniformly distributed in the region [0, 2π] as:

$$Phase\ Mask = \exp(i2\pi\varnothing(v,u)) \tag{7}$$

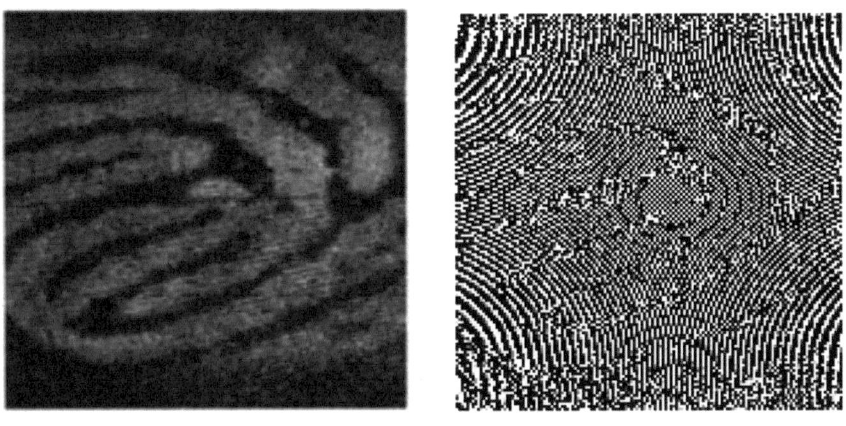

(a) (b)

Figure 3.
Reconstructed features: (a) amplitude distribution and (b) phase distribution.

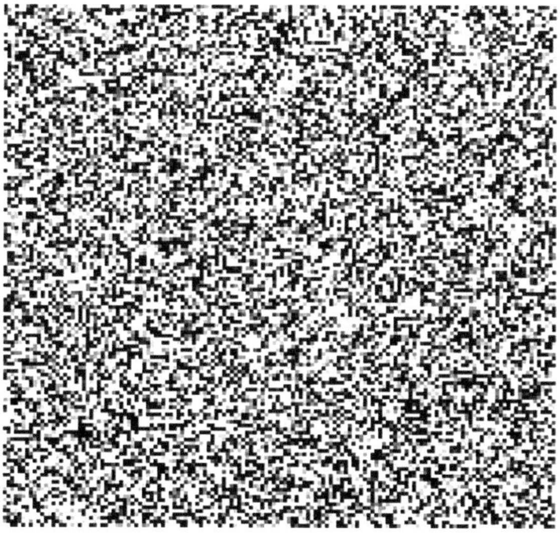

Figure 4.
Generated fingerprint phase mask (PM).

The generated phase mask is full of speckle patterns and randomness, as shown in **Figure 4**. This is also unique to the person because of the associated fingerprint biometrics.

Our previous works have shown that the biometric keys from fingerprint hologram are a promising candidate for image encryption. A detailed description of the biometric key characteristics, such as uniqueness, randomness, and robustness, is recently reported in our published work [32–34, 57]. Motivated by the utility of biometric keys, the authors presented a cryptosystem for image encryption and decryption.

3. Basic cryptography in the perspective of biometric authentication

In the domain of data security, the public key cryptographic technique is used to secure the information using the public keys (K_{Public}) and the private keys ($K_{Private}$) [53]. The cryptographic process between Alice and Bob is shown in **Figure 5**, which can be explained as:

1. In the cryptographic process, the public key is utilized as the key for encryption while the private key is used only for decryption.

2. To transfer the ciphertext to Alice, Bob applies the use of a public key for encrypting the input data.

3. At the decryption stage, Alice involves the use of private keys to get back the input information from the ciphertext using the decryption process. Moreover, Alice cannot decrypt the unspecified ciphertext because of the nonavailability of their private keys.

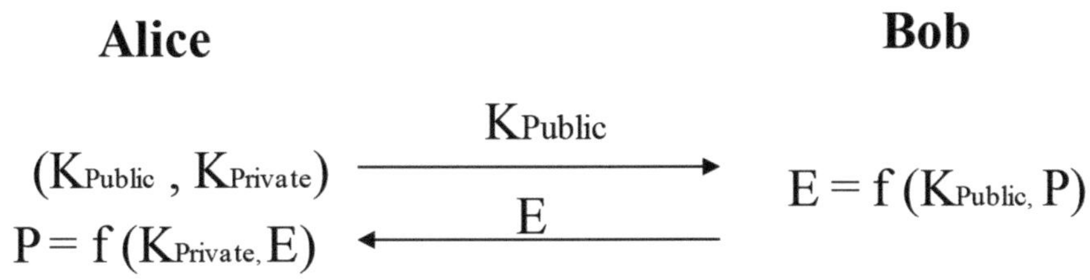

Figure 5.
Basic public key cryptography for encryption: P = plaintext, E = encrypted or ciphertext, and f = processing algorithm.

The purpose of the cryptographic process is to convert plaintext information into ciphertext. Moreover, its security strength depends on keys and processing algorithms which are not linked with the user's identity [53]. In general, security systems use token, ID, and password to authenticate the person but still suffer from several issues and limitations with regard to information security as well as insufficient database to prevent unauthorized access [5]. In order to implement a biometric-based authentication approach, the biometrics of a person must be first registered into the system that works only for an authorized person and it would not work if a person is not registered. For this purpose, a fingerprint hologram of a person using a digital holographic process is utilized, as shown in **Figure 6**. The digital process for recording, as well as retrieval of the keys, makes it safe, secure, and accessible for encryption and decryption processes. In addition, the use of experimental parameters results in additional security for the system [32–34, 57].

In view of cryptography using a biometric perspective, Alice uses her biometric keys retrieved from fingerprint hologram for encoding the plaintext information, and Bob confirms Alice's biometric keys by matching them to the registered database so that it can assure the ciphertext coded by Alice [33]. In this way, this strategy satisfies the criteria of asymmetric cryptography with authentication. The rules of the system can be elaborated as:

1. If Bob wishes to transmit the input information to Alice. In this case, Bob first needs to register his fingerprint biometric details using the process of digital holography.

2. In the next step, Bob employs the encryption algorithm for converting the information into the ciphertext using the public key by including biometric keys.

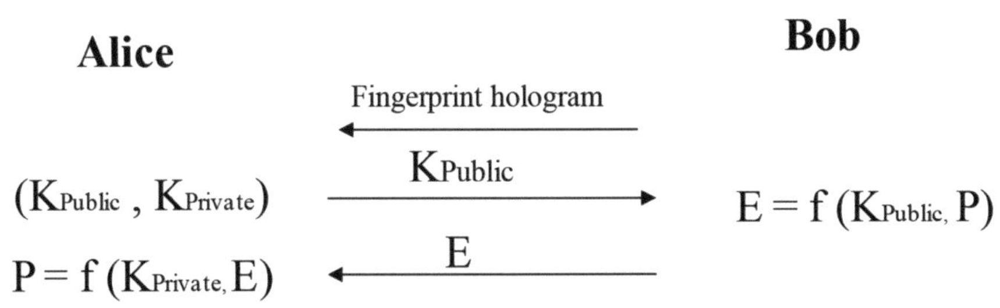

Figure 6.
Cryptography perspective using biometrics.

3. In order to decode, Alice needs the simultaneous presence of the private keys and the availability of Bob's fingerprint hologram to obtain biometric keys.

Hence, the main contribution of this chapter is to present an optical nonlinear cryptosystem by involving biometric authentication using a fingerprint hologram. The authors evaluated measures of the optical cryptosystem process in achieving keys in terms of public and private keys for encryption and decryption with higher levels of security.

4. Optical cryptosystem using biometric keys

This section presents algorithms for encoding of the image by the use of biometric keys.

4.1 Encryption process

To encrypt the input information using the optical cryptosystem, a preprocessing layer of the phase-retrieval algorithm using biometric keys generated from fingerprint hologram has been included, and then the PTFT scheme is employed. The flow diagram of the proposed encryption system is shown in **Figure 7**.

In order to do this, the input image (I) is initially encoded by involving the biometric keys from the fingerprint hologram. This encoding applies constraints as fingerprint amplitude by replacing the amplitude in the Fourier domain, while the Fourier phase is kept unchanged [31]. This process implements iteratively back and forth between the object domain and the Fourier domain. The iteration number is

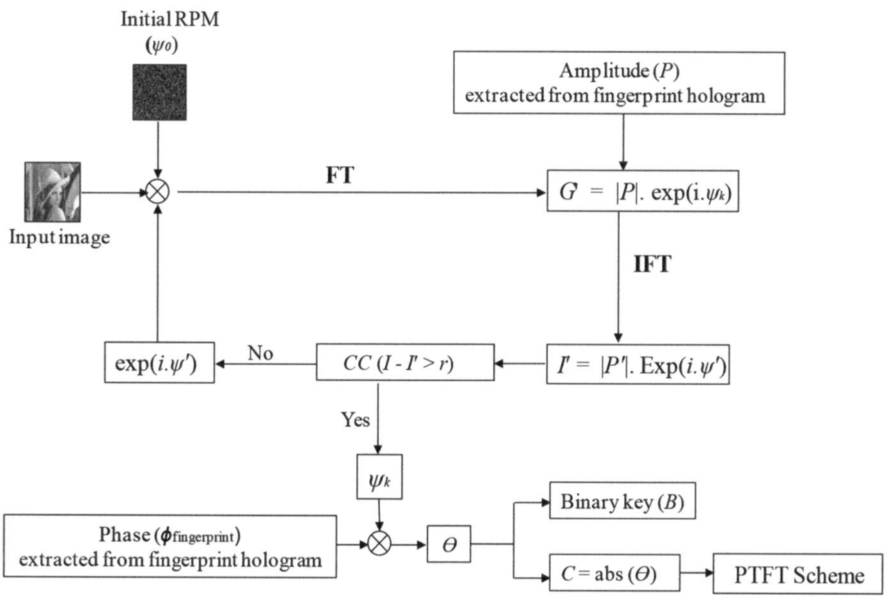

Figure 7.
Flow diagram of the proposed encryption scheme.

decided between the input image (I) and the retrieved image (I') by measuring the correlation coefficient (CC) as:

$$CC = \frac{\sum_{x=1}^{M}\sum_{y=1}^{N}(I(x,y) - \bar{I})\left(I'(x,y) - \bar{I'}\right)}{\sqrt{\sum_{x=1}^{M}\sum_{y=1}^{N}(I(x,y) - \bar{I})^2}\sqrt{\sum_{x=1}^{M}\sum_{y=1}^{N}\left(I'(x,y) - \bar{I'}\right)^2}} \tag{8}$$

As shown in Eq. (8), the average values of the input image ($I(x,y)$) and the retrieved image ($I'(x,y)$) are represented as $\bar{I}$ and $\bar{I'}$, respectively. The domain (x,y) represents information in the image plane, while M and N show the row and column of the image. The CC values are plotted with the number of iterations that illustrate the error between the decrypted image and the input image continuously keeps going down as the number of iterations increases. This process facilitates the retrieval of better-quality images, as shown in **Figure 8**.

From the retrieved image as shown in **Figure 8**, when the CC values reach the desired level (≥ 0.998) then the iteration process is stopped and its output (ψ_k) is combined with fingerprint phase ($\phi_{fingerprint}$) (as)

$$\psi_k \otimes \phi_{fingerprint} = \Theta \tag{9}$$

Here, $\otimes$ shows the multiplication operator. This resultant information (Θ) is further distributed into two parts:

(a) Binary key (B): For binary key (B), the resultant information (Θ) is coded using the following mathematical identity:

$$B = \begin{cases} 0, & \Theta < 0 \\ 1, & \Theta \geq 0 \end{cases} \tag{10}$$

If Θ is greater than or equal to zero then it is set to be 1 while if Θ is less than zero, it is denoted by zero. This result is coined as a binary key and kept as the private key, which is protected using the pixel scrambling operation to make it safe for transmission [33].

(b) C = abs(Θ): The absolute information is just an intensity distribution that looks like a speckle, in which the fingerprint biometric features are deeply hidden.

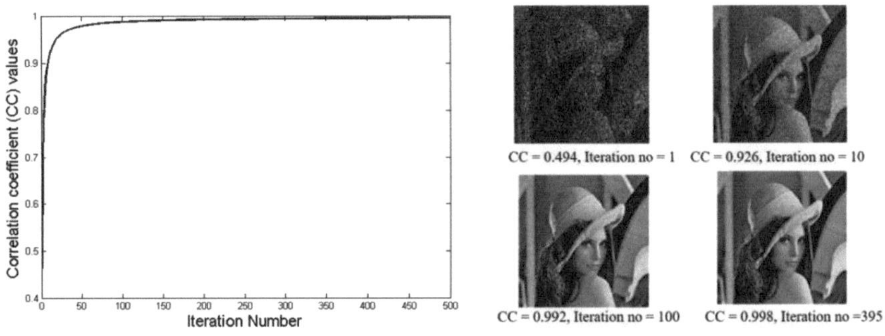

(a) (b)

Figure 8
(a) Graph between the number of iterations and correlation coefficient (CC) and (b) retrieved image

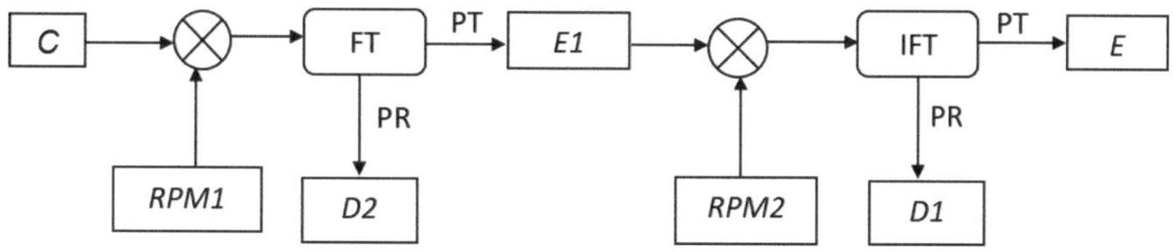

Figure 9.
Phase-truncated Fourier transform (PTFT) scheme for encryption.

Moreover, to obtain the ciphertext (E), the information 'C' is processed using the two encryption keys ($RPM1$ and $RPM2$) at the input plane and the Fourier plane using the PTFT operation, as shown in **Figure 9**. This process is expressed as:

$$E1 = PT\{FT\,[C \cdot RPM1]\} \tag{11}$$

$$E = PT\{IFT\,[E1 \cdot RPM2]\} \tag{12}$$

From the process, the two decryption keys ($D1$ and $D2$) are obtained as:

$$D2 = PR\{FT\,[C \cdot RPM1]\} \tag{13}$$

$$D1 = PR\{IFT\,[E1 \cdot RPM2]\} \tag{14}$$

From the reported process, the three keys are obtained during encryption processes, which are kept as the private keys to decode the information. As a result, our system involves the use of public keys to encode the input data that can only be decoded using the private keys while the fingerprint keys corroborate the user specificity throughout the optical encryption and decryption processes.

4.2 Decryption process

In order to retrieve the information, the fingerprint hologram and the reconstruction parameters are provided to be known to the user, which further enables the process to recover the original information. This is only possible if both the provided information is correct. Therefore, this processing step has the capability to confirm the authenticity of the person. First, the decryption process is performed to retrieve the information (C) from the ciphertext (E), as shown in **Figure 10**.

$$E1 = PT\{FT[E \cdot D1]\} \tag{15}$$

$$C = PT\{IFT\,[E1 \cdot D2]\} \tag{16}$$

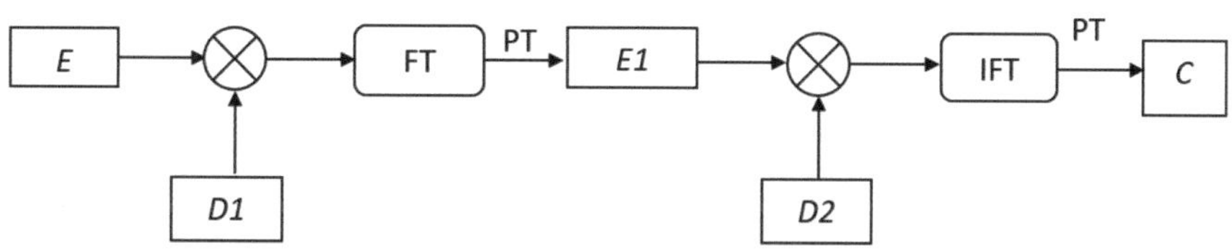

Figure 10.
Phase-truncated Fourier transform (PTFT) scheme for decryption.

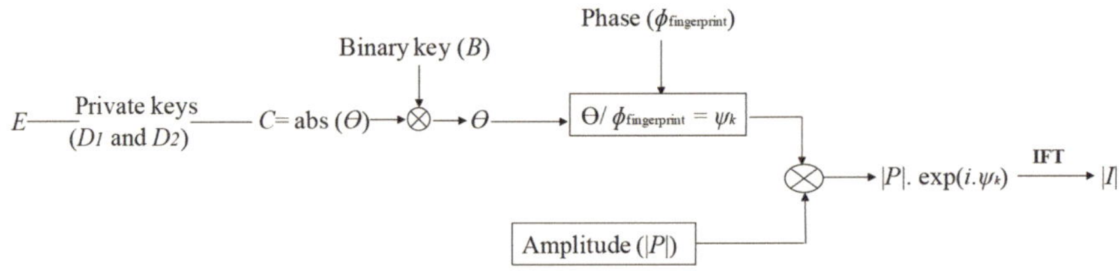

Figure 11.
Flow diagram of the decryption process.

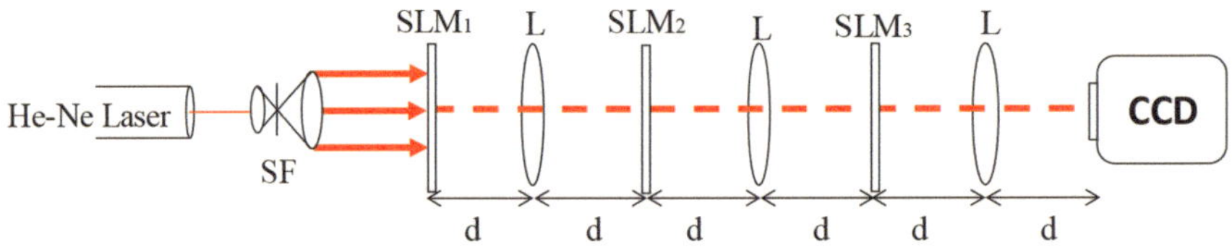

Figure 12.
Optical setup for decryption: CCD: charge-coupled device, SLM: spatial light modulator, d: focal length of lenses, and SF: spatial filtering.

For decryption, the initially encoded information (Θ) is first obtained from the ciphertext (C) by applying the binary key and the scrambling key. As shown in **Figure 11**, the biometric keys as the phase $\left(\Phi_{Fingerprint}\right)$ data and the magnitude (P) are involved to obtain the input image (I). These decryption steps are mathematically given by

$$\Theta = C \otimes B = abs(\Theta) \otimes B \tag{17}$$

$$\psi_k = \Theta / \Phi_{Fingerprint} \tag{18}$$

$$I = IFT(|P| \exp(i\psi_k)) \tag{19}$$

where, the term 'abs' represents the absolute value of a matrix.

Figure 12 shows the implementation of the optical experimental setup for decrypting the information using electronic devices such as spatial light modulators (SLMs) and CCD, which are controlled by a personal computer. At the beginning of the process, the combined data of the encrypted information (E) with the private key (D1) are shown in the SLM_1 device and then illuminated by the He-Ne laser beam to carry out optical Fourier transform (FT). This result combined with the second private key (D2) is shown in the SLM_2 device, which is further optical FT. In the last step, the obtained information is involved digitally using the binary key (B) and biometric keys. By performing optical FT, the CCD captures the decrypted information.

5. Results and discussion

This section evaluates the performance of our system by performing a number of computer simulations on a MATLAB platform. The obtained results of the system validate the effectiveness of our scheme that exhibits higher security with keys management and distribution.

5.1 Input data

In this section, the computer simulation results of our system are presented. In our experiments, the size of all images employed is 146 × 146 pixels. The input image to be encrypted and the generated biometric keys from the fingerprint hologram are shown in **Figure 13**a–c. The RPM keys for the PTFT scheme are shown in **Figure 13**d–e. Our system obtains ciphertext, as shown in **Figure 13**f.

5.2 Decryption results

To evaluate our system, a series of decryption experiments are carried out. In the first experiments, we wished to recover the input information against unauthorized attempts using the possible key combinations in **Figure 14**a–e. Simulation results indicate that **Figure 14a** shows the truly decrypted input information using all keys in the correct order with authentication. **Figure 14b** shows the recovered image using the private keys (D1 and D2) in the wrong positions. **Figure 14c** represents the recovered information when no keys are employed for decryption. **Figure 14d** displays the obtained image by applying any arbitrarily generated binary key. **Figure 14e** represents the decrypted noisy data when the biometric phase key is wrongly employed. In addition, to investigate the quality of the recovered data, the mean-square error (MSE) measures for **Figure 14**a–e are evaluated as 1.5740×10^{-4}, 0.2132, 0.1313, 0.1163, and 0.2609, respectively. Moreover, the CC parameter between the

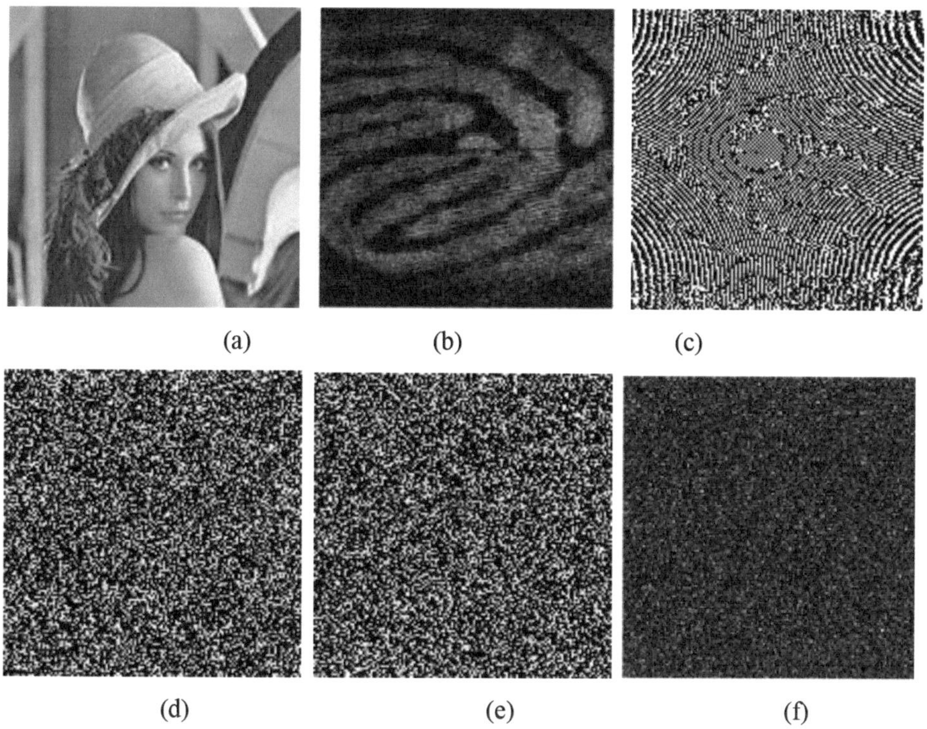

Figure 13.
(a) Input image. (b) Fingerprint AM key. (c) Fingerprint phase key. (d) RPM1 key. (e) RPM2 key. (f) Ciphertext (E).

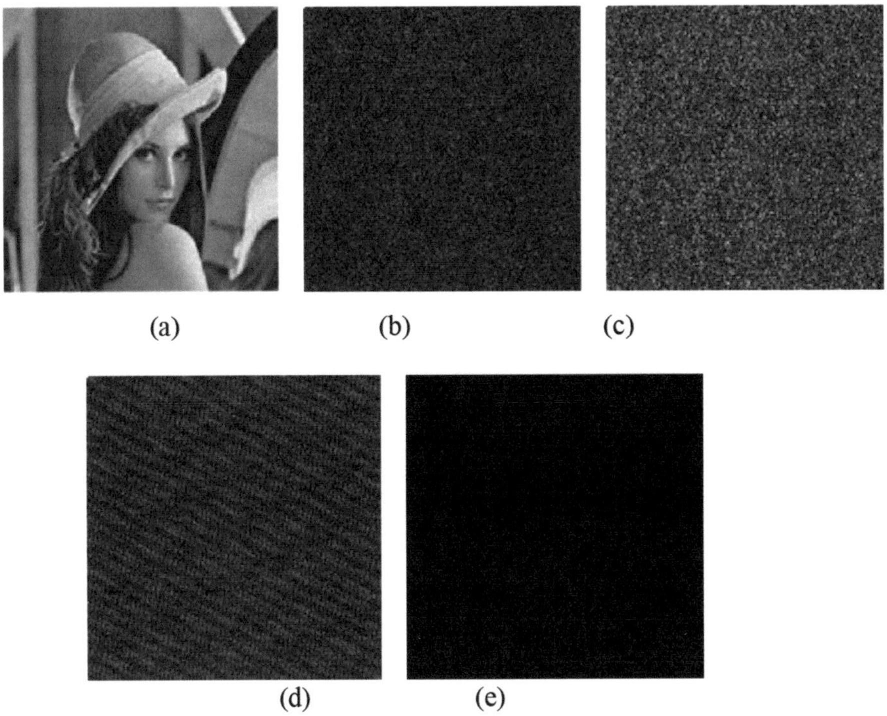

Figure 14.
Decryption results using (a) all correct keys, (b) keys (D1 and D2) in wrong positions, (c) no keys, (d) wrong binary key, and (e) different fingerprint phase keys.

input image as shown in **Figure 13a**, and the retrieved image as displayed in **Figure 14a** is calculated as a value of 0.998.

5.3 Effect of quantization

In the next step of experiments to make digital simulations closer to the true physical process, we evaluated the influence of the quantization at different quantization levels in the process [33]. For this purpose, the encrypted data shown in the SLM device and the retrieved information were captured by a CCD camera that is quantized at different levels of quantization such as 5 bits, 8 bits, 10 bits, 12 bits, and 16 bits, respectively. In this context, the obtained images are displayed in **Figure 15**. To examine the effect of quantization, the CC and MSE values are calculated. As shown in **Table 1**, the results illustrate the high accuracy of our system.

5.4 Robustness of the binary key

In this section, we evaluated the system performance against the binary key. As explained in Section 3.2, the binary key is produced as one of the private keys that are secured by pixel scrambling operation. Using this approach, if the attacker knows the total number of pixels present in the key, this is not sufficient to know the exact distribution of zeros or ones. Keeping in view this point, an attempt is performed using the true binary values of the key and its wrong distribution. We have illustrated results as shown in **Figure 16a** and **b** by plotting MSE and CC values between the

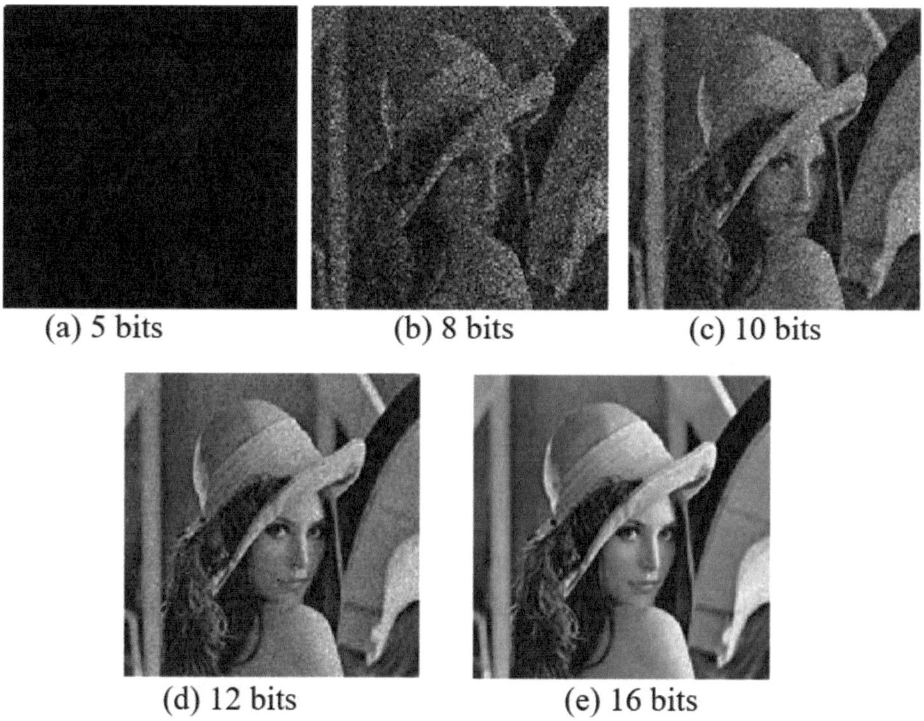

Figure 15.
Decrypted image for different quantization levels: (a) 5 bits, (b) 8 bits, (c) 10 bits, (d) 12 bits, and (e) 16 bits.

Quantization level	MSE value	CC value
5 bits	0.2276	0.3525
8bits	0.0654	0.6527
10 bits	0.0240	0.8812
12 bits	0.0045	0.9793
16 bits	2.6235×10^{-4}	0.9968

Table 1.
Showing performance at different quantization levels.

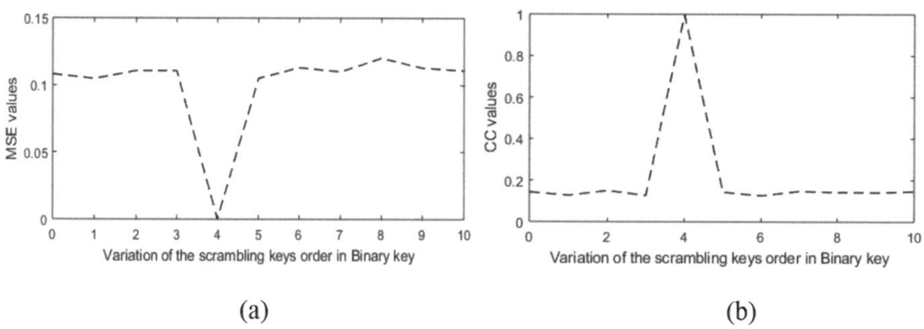

Figure 16.
Graph to illustrate robustness of the binary key (a) mean-square error (MSE) curve and (b) correlation coefficient (CC) curve.

original image and decrypted images. Experimental results indicate that the true input image is recovered if the binary key is correct while other retrieved images using the wrong binary key represent noisy information.

5.5 Security analysis and discussion

In this section, we evaluated the system performance against iterative phase-retrieval algorithms as reported by researchers [16, 19]. In our work, the fingerprint object information about the amplitude and the phase keys is exploited for encrypting the input information. In this context, the inclusions of the biometric keys make the encryption process relevant to be user-specific. This implication breaks the linearity and enhances the nonlinearity and complexity of the encryption process [32, 33]. In order to access the image using the attack algorithm, the inherent noise goes on boosting at each iterative step. Thus, our system provides resistance against the attacks. In order to prove this point, cryptanalysis was conducted against the attacks such as the special attack and known-plaintext attack (KPA). To illustrate the attack process the special attack is used to break the cryptosystem with the two encryption keys RPMs and the ciphertext, which are allowed to be known to the attacker. This attack is based on a two-step iterative algorithm as mentioned in [16, 19]. Using known resources as shown in **Figure 13**, the attacker retrieved the information as shown in **Figure 17a**. Moreover, in a similar manner, our system performed cryptanalysis for the KPA whose results are shown in **Figure 17b**. Thus, our systems indicate the robustness of the proposed scheme against both the special attack and the KPA.

5.6 Comparison with other schemes

Finally, we present a comprehensive comparative performance of our system with the recently published schemes [20, 26, 27, 30, 31, 40, 43–46]. Alarifi *et al.* [20] introduced an optical PTFT-based asymmetric encryption algorithm for biometric template protection using a cancellable approach. Takeda *et al.* [26] reported a smart card holder authentication based on the DRPE scheme, in which the encryption key as

(a) (b)

Figure 17.
Attack results: (a) special attack. (b) Known-plaintext attack (KPA).

Fourier phase data of the fingerprint is involved, which gives rise to issues of wrong authentication due to positional variation of fingerprint at the enrollment and verification stages. Saini *et al.* [27] worked on optical security using a DRPE system that uses encryption keys linked to the biometrics of a user which offers a solution to keys distribution. Tashima *et al.* [30] presented an improved DRPE security by avoiding the known-plaintext attack. Mehra *et al.* [43] reported recently an asymmetric system for encrypting the fingerprint image based on quick response (QR) decomposition in the domain of gyrator wavelet transform. Castro *et al.* [45] proposed an encryption scheme for medical images based on fingerprint authentication. Souza *et al.* [44] reported an optical encryption technique in which passwords and tokens were included as multifactor for authentication. Chang *et al.* [46] developed an asymmetric encryption scheme using optical scanning cryptography by combining elliptic curve cryptography, which also helps to achieve keys management. In comparison with the reported encryption algorithms for information using biometrics as reported in [20, 26, 27, 30, 31, 40, 43–46], our scheme uses the biometric keys obtained from the fingerprint hologram, which is protected by experimental parameters, which help to enable verification and authentication at decryption stage. The digital approach of biometric key generation is safe and secure for encryption and decryption processes. Based on the results obtained from our system, it can resist several types of potential attacks because of its complexity, nonlinearity, and robustness in comparison to other reported cryptosystems. The obtained outcomes in **Table 2** demonstrated that the security performance measures of our system are superior and reliable as compared to those mentioned in the previously published work. Our system has a simple experimental implementation that has included the involvement of optoelectronics components and devices. Thus, our system is efficient in terms of security including the distribution of keys with information authentication.

Authors	Methods	Security performance measures
Alarifi et al. [20]	Optical PTFT Asymmetric Cryptosystem-Based Secure and Efficient Cancelable Biometric Recognition System.	Probability of false distribution, equal error rate, false rejection ratio, false acceptance ratio, correlation analysis
Takeda et al. [26]	Encoding plaintext by Fourier transform hologram in double random phase encoding using fingerprint keys	Equal error rate, false rejection ratio, false acceptance ratio
Saini et al. [27]	Biometrics-based key management of double random phase encoding scheme using error control codes	Equal error rate attack-resistant, key management
Tashima et al. [30]	Known-plaintext attack on double random phase encoding using fingerprint as key and a method for avoiding the attack	Sum-squared error, phase-only correlation, known-plaintext attack
Zhu et al. [31]	Computational ghost imaging encryption based on fingerprint phase mask	Mean-square error, correlation coefficient, robust against iterative algorithm attacks
Zhao et al. [40]	Image encryption using fingerprint as key based on phase-retrieval algorithm and public key cryptography	Mean-square error, correlation coefficient, robust against iterative algorithm attacks, fingerprint authentication
Mehra et al. [43]	Fingerprint image encryption using phase-retrieval algorithm in gyrator wavelet transform domain using QR decomposition	Brute force attack, known-plaintext attack, and special attack

Authors	Methods	Security performance measures
Castro et al. [44]	A Medical Image Encryption Scheme for Secure Fingerprint-Based Authenticated Transmission	Mean-square error, peak signal-to-noise ratio, running efficiency, authentication
Souza et al. [45]	Improving biometrics authentication with a multifactor approach based on optical interference and chaotic maps	mean-square error, correlation coefficient, entropy, authentication
Chang et al. [46]	Asymmetric cryptosystem based on optical scanning cryptography and elliptic curve algorithm	Key management, authentication, robust against attacks
Our work [32, 33, 57]	Biometric-based optical systems for security and authentication	Equal error rate, false rejection ratio, false acceptance ratio, mean-square error, correlation coefficient Known-plaintext attack, special attack, key management, authentication

Table 2.
Comparative study of our work with recent optical cryptosystems.

6. Conclusions and future work

In this chapter, we have described the optical nonlinear cryptosystem using fingerprint biometric keys based on phase retrieval and the PTFT scheme for image encryption. It also showed how fingerprint biometrics can be captured based on optical implementation using digital holography in a practical manner without inconvenience to the person. Our system has the salient features that the capability of biometric key retrieval from fingerprint hologram is led to authenticate the person who possesses the ciphertext for decrypting the information. In addition, our system meets the criteria of asymmetric encryption approach which help to provide a solution for the key management and distribution in the encryption and decryption processes. This system has simple implementation either numerically or optically. As a result, we could mention clearly that our system has validity and robustness against unauthorized attempts and well-known attacks.

In future work, a study about the use of the optoelectronic system to record a hologram of the real fingerprint pattern of a person will be performed in order to explore the scientific potential of the emerging field and make clear the validity of our cryptosystem by evaluating security performance.

Acknowledgements

This work was supported by the National Natural Science Foundation of China (61875129 and 62061136005).

Conflict of interest

The authors declare no conflict of interest.

Author details

Gaurav Verma[1]*, Wenqi He[2] and Xiang Peng[2]

1 Department of Electronics and Communication Engineering, B.K. Birla Institute of Engineering and Technology, Pilani, Rajasthan, India

2 College of Physics and Optoelectronic Engineering, Shenzhen University, Shenzhen, P.R. China

*Address all correspondence to: gaurav.sgs85@gmail.com

References

[1] Jain AK, Ross A, Pankanti S. Biometrics: A tool for information security. IEEE Transactions on Information Forensics and Security. 2006;**1**:125-143

[2] Wayman JL. Fundamentals of biometric authentication technologies. International Journall of Image Graph. 2001;**1**:93-113

[3] Ratha NK, Connell JH, Bolle RM. Enhancing security and privacy in biometrics-based authentication systems. IBM Systems Journal. 2001;**40**:614-634

[4] Jain AK, Flynn P, Ross A. Handbook of Biometrics. 1st ed. New York, NY, USA: Springer; 2008

[5] Javidi B. Optical and Digital Techniques for Information Security. 1st ed. New York: Springer; 2005

[6] Liu S, Guo C, Sheridan JT. A review of optical image encryption techniques. Optics and Laser Technology. 2014;**57**: 327-342

[7] Chen W, Javidi B, Chen X. Advances in optical security systems. Advances in Optics and Photonics. 2014; **6**:120-155

[8] Refregier P, Javidi B. Optical image encryption based on input plane encoding and Fourier plane random encoding. Optics Letters. 1995;**20**: 767-769

[9] Unnikrishnan G, Joseph J, Singh K. Optical encryption by double-random phase encoding in the fractional Fourier domain. Optics Letters. 2000;**25**:887-889

[10] Chen L, Zhao D. Optical image encryption with Hartley transforms. Optics Letters. 2006;**31**:3438-3440

[11] Situ G, Zhang J. Double random-phase encoding in the Fresnel domain. Optics Letters. 2004;**29**:1584-1586

[12] Peng X, Zhang P, Wei H, Yu B. Known-plaintext attack on optical encryption based on double random phase keys. Optics Letters. 2006;**31**: 1044-1046

[13] Peng X, Wei H, Zhang P. Chosen-plaintext attack on lens less double-random phase encoding in the Fresnel domain. Optics Letters. 2006;**31**: 3261-3263

[14] Qin W, Peng X. Asymmetric cryptosystem based on phase truncated Fourier transforms. Optics Letters. 2010; **35**:118-120

[15] Qin W, Peng X, Meng X, Gao B. Universal and special keys based on phase-truncated Fourier transform. Optical Engineering. 2011;**50**:080501

[16] Wang X, Zhao D. A special attack on the asymmetric cryptosystem based on phase-truncated fractional Fourier transforms. Optics Communication. 2012;**285**:1078-1081

[17] He W, Pan S, Liao M, Lu D, Xing Q, Peng X. A learning-based method of attack on optical asymmetric cryptosystems. Optics and Lasers in Engineering. 2021;**138**:106415

[18] Pan S, Liao M, He W, Zhang Y, Peng X. Untrained neural network for cryptanalysis of a phase-truncated-Fourier-transform-based optical cryptosystem. Optics Express. 2021;**29**(26):42642-42649

[19] Rajput SK, Nishchal NK. Known-plaintext attack-based optical

cryptosystem using phase-truncated Fresnel transform. Applied Optics. 2013; **52**:871-878

[20] Alarifi A, Amoon M, Aly MH, El-Shafai W. Optical PTFT asymmetric cryptosystem-based secure and efficient Cancelable biometric recognition system. IEEE Access. 2020;**8**: 221246-221268

[21] Sinha A. Nonlinear optical c ryptosystem resistant to standard and h ybrid attacks. Optics and Lasers in Engineering. 2016;**81**:79-86

[22] Rajput SK, Nishchal NK. Optical double image security using random phase fractional Fourier domain encoding and phase-retrieval algorithm. Optics Communication. 2017;**388**:38-46

[23] Xiong Y, He A, Quan C. Specific attack and security enhancement to optical image cryptosystem based on two random masks and interference. Optics and Lasers in Engineering. 2018;**107**: 142-148

[24] Jiao S, Zhuang Z, Zhou C, Zou W, Li X. Security enhancement of double random phase encryption with a hidden key against ciphertext only attack. Optics Communication. 2018;**418**: 106-114

[25] Verma G, Sinha A. Optical image encryption system using nonlinear approach based on biometric authentication. Journal of Modern Optics. 2017;**64**:1321-1329

[26] Takeda M, Nakano K, Suzuki H, Yamaguchi M. Encoding plaintext by Fourier transform hologram in double random phase encoding using fingerprint keys. Journal of Optics. 2012;**14**:094003

[27] Saini N, Sinha A. Biometrics based key management of double random phase encoding scheme using error control codes. Optics and Lasers in Engineering. 2013;**51**:1014-1022

[28] Yan A, Wei Y, Zhang J. Security enhancement of optical encryption based on biometric array keys. Optics Communication. 2018;**419**:134-140

[29] Suzuki H, Yamaguchi M, Yachida M, Ohyama N, Tashima H, Obi T. Experimental evaluation of fingerprint verification system based on double random phase encoding. Optics Express. 2006;**14**:1755-1766

[30] Tashima H, Takeda M, Suzuki H, Obi T, Yamaguchi M, Ohyama N. Known plaintext attack on double random phase encoding using fingerprint as key and a method for avoiding the attack. Optics Express. 2010;**18**:13772-13781

[31] Zhu J, Yang X, Meng X, Wang Y, Yin Y, Sun X, et al. Computational ghost imaging encryption based on fingerprint phase mask. Optics Communication. 2018;**420**:34-39

[32] Verma G, Sinha A. Securing information using optically generated biometric keys. Journal of Optics. 2016; **18**:115701

[33] Verma G, Liao M, Lu D, He W, Peng X, Sinha A. An optical asymmetric encryption scheme with biometric keys. Optics and Lasers in Engineering. 2019; **116**:32-40

[34] Verma G, He W, Peng X. A novel four image encryption approach in sparse domain based on biometric keys. Multimedia Tools and Applications. 2023;**82**:22889-22904. DOI: 10.1007/ s11042-023-14801-7

[35] Verma G, He W, Lu D, Liao M, Peng X, Healy J, et al. Securing multiple

information using bio-chaotic keys. IEEE Photonics Journal. 2021;**13**(1):1-17

[36] Xiong Y, Gu J, Kumar R. Security analysis on an interference-based optical image encryption scheme. Applied Optics. 2022;**61**:9045-9051

[37] He W, Peng X, Meng X. Optical multiple-image hiding based on interference and grating modulation. Journal of Optics. 2012;**14**: 075401

[38] Chen J, He W. Parallel optical hash function based on the interaction between linearly polarized light and multiple-scattering media. Applied Optics. 2022;**61**:5457-5465

[39] Rajput SK, Nishchal NK. Fresnel domain nonlinear optical image encryption scheme based on Gerchberg–Saxton phase-retrieval algorithm. Applied Optics. 2014;**53**:418-425

[40] Zhao T, Ran Q, Yuan L, Chi Y, Ma J. Image encryption using fingerprint as key based on phase retrieval algorithm and public key cryptography. Optics and Lasers in Engineering. 2015;**72**:12-17

[41] Zhao T, Ran Q, Yuan L, Chi Y, Ma J. Optical image encryption using password key based on phase retrieval algorithm. Journal of Modern Optics. 2016;**63**:771-776

[42] Zhao T, Ran Q, Chi Y. Image encryption based on nonlinear encryption system and public key cryptography. Optics Communication. 2015;**338**:64-72

[43] Mehra I, Nishchal NK. Fingerprint image encryption using phase retrieval algorithm in gyrator wavelet transform domain using QR decomposition. Optics Communication. 2023;**533**:129265

[44] Castro F, Impedovo D, Pirlo G. A medical image encryption scheme for secure fingerprint – Based authenticated transmission. Applied Sciences. 2023;**13**: 6099

[45] Souza D, Burlamaqui A, Filho GS. Improving biometrics authentication with a multi-factor approach based on optical interference and chaotic maps. Multimedia Tools and Applications. 2018;**77**:2013-2032

[46] Chang X, Li W, Yan A, Tsang PWM, Poon TC. Asymmetric cryptosystem based on optical scanning cryptography and elliptic curve algorithm. Scientific Reports. 2022;**12**:7722

[47] Lu D, Liao M, He W, Xing Q, Verma G, Peng X. Experimental optical secret sharing via an iterative phase retrieval algorithm. Optics and Lasers in Engineering. 2020;**126**:105904

[48] Liu W, Liu Z, Liu S. Asymmetric cryptosystem using random binary phase modulation based on mixture retrieval type of Yang-Gu algorithm. Optics Letters. 2013;**38**:1651-1653

[49] Liu W, Liu Z, Liu S. Asymmetric cryptosystem using random binary phase modulation based on mixture retrieval type of Yang-Gu algorithm: Reply. Optics Letters. 2013;**38**:4045

[50] He W, Meng XF, Peng X. Asymmetric cryptosystem using random binary phase modulation based on mixture retrieval type of Yang-Gu algorithm: Comment. Optics Letters. 2013;**38**(20):4044

[51] Wang X, Chen W, Chen X. Optical information authentication using compressed double-random-phase-encoded images and quick-response codes. Optics Express. 2015;**23**:6239-6253

[52] Chen H, Tanougast C, Liu Z, Sieler L, Ramenah H. Optical image asymmetric cryptosystem using fingerprint based on iterative fraction Fourier transform. Optical and Quantum Electronics. 2017; **49**:157

[53] Stallings W. Cryptography and Network Security: Principles and Practice. 2nd ed. Upper Saddle River, NJ, United States: Prentice Hall; 1999

[54] Schnars U, Juptner W. Direct recording of holograms by a CCD-target and numerical reconstruction. Applied Optics. 1994;**33**:179-181

[55] Schanrs U, Juptner W. Digital recording and numerical reconstruction of holograms. Measurement Science and Technology. 2002;**13**:R85-R101

[56] Javidi B, Takanori N. Securing information by use of digital holography. Optics Letters. 2000;**25**:28-30

[57] Verma G, Sinha A. Digital holographic-based cancellable biometric for personal authentication. Journal of Optics. 2016;**18**:055705

[58] Liao M, He W, Lu D, Wu J, Peng X. Security enhancement of the phase-shifting interferometry-based cryptosystem by independent random phase modulation in each exposure. Optics and Lasers in Engineering. 2017; **89**:34-39

[59] Potcoava MC, Kim MK. Fingerprint biometry applications of digital holography and low-coherence interferography. Applied Optics. 2009; **48**:H9-H15

[60] Lu D, Xing Q, Liao M, Situ G, Peng X, He W. Single-shot noninvasive imaging through scattering medium under white light illumination. Optics Letters. 2022;**47**(7):1754-1757

[61] Liao M, Feng Y, Lu D, Li X, Giancarlo P, Karsten F, et al. Scattering imaging as a noise removal in digital holography by using deep learning. New Journal of Physics. 2022;**24**:083014

[62] He W, Lu WY, D, Li X, Liao M, Peng X. Noninvasive imaging of two isolated objects through a thin scattering medium beyond the 3D optical memory effect by speckle-based difference strategy. Optics Letters. 2021;**46**(23): 5954-5957

[63] Larkin KG, Fletcher PA. A coherent framework for fingerprint analysis: Are fingerprints holograms? Optics Express. 2007;**15**:8667

[64] Zhang C, Han B, He W, Peng X, Xu C. A novel compressive optical encryption via single-pixel imaging. IEEE Photonics Journal. 2019;**11**:7801208

[65] Cheng Y, Larin KV. Artificial fingerprint recognition by using optical coherence tomography with autocorreclation analysis. Applied Optics. 2006;**45**:9238-9245

[66] FVC 2004 Data Base. http://bias.csr.unibo.it/fvc2004/databases.asp

Chapter 10

Secure Smart Card IP

El Hadj Youssef Wajih

Abstract

This book chapter highlights the embedded system security by designing a secure smart card IP. Indeed, the smart card is recognized as a privileged means of both storing confidential information and performing secure transactions. Its main role comes from the security it provides inside the system it is a part of. The specification and development of the elaborate smart card architecture are very delicate steps that require the pooling of strong competences in computer security, electronics, and also cryptography. The developed secure smart card IP model is based on the Gaisler LEON2 processor. To ensure a maximum level of security and optimal performance, a hardware integration of cryptographic mechanisms through instruction extensions was carried out. The integrated mechanisms allow for ensuring confidentiality, hashing, random number generation, and digital signature. The proposed smart card IP was implemented on a reconfigurable FPGA platform, and then on ASIC using 40 nm CMOS technology. A surface area of 1.08 mm^2 with a consumed dynamic power of 23 mW for a frequency of 13.5 MHz was achieved.

Keywords: cryptography, smart card, Leon2 processor, FPGA, ASIC

1. Introduction

Smart cards, as embedded systems utilized by consumers, play a crucial role in safeguarding the security of their respective systems. However, the potential of smart cards has significantly expanded with the introduction of multi-application cards, offering a diverse range of services such as GSM, electronic wallets, and loyalty programs [1]. To ensure confidentiality, security, and authentication, the integration of cryptographic mechanisms into smart cards is of utmost importance.

With the increasing diversity and openness of smart card systems, a race ensues between smart card developers and attackers, aiming to discover vulnerabilities and bolster security measures. Consequently, it becomes imperative to continuously update smart card security in order to counter hardware attacks effectively. This necessitates the implementation of robust countermeasures capable of detecting and thwarting attempts to manipulate the card's behavior or exploit techniques like spatial, temporal, or information redundancy [2].

The central objective of this chapter is to design a secure smart card Intellectual Property (IP). This endeavor encompasses the careful selection of appropriate hardware components, comprising essential blocks, and the development of an efficient interconnection system. Throughout the design process, thorough consideration is given to performance criteria, adherence to industry standards, specific characteristics pertinent to smart cards, and any applicable constraints. Each individual block of the design will undergo meticulous evaluation at multiple stages of the design chain, thereby ensuring the integrity and efficacy of the overall system.

2. State of the art and objectives

A smart card is a plastic card that contains an electronic circuit capable of securely manipulating information, such as storing and calculating. The evolution of smart card technology has been marked by various significant dates. In 1974, Roland Moreno, leading a research team for Innovation, created the first memory-based smart cards. In 1977, the memory card advanced into a microprocessor card. In 1980, the French company Bull produced the CP8, the first microprocessor card used for early trials of bank cards [3]. By 1984, the first health smart card was introduced, and the micro-module card emerged in the same year to create the first telephone cards. In 1996, the publication of the Java Card 1.0 specification by Schlumberger simplified smart card programming. The following year, Bull, Sun, and Gemplus collaborated with Schlumberger to found the Java Card Forum, marking the beginning of smart card standards and specifications [4].

3. Application areas

The fields of application of smart cards have continued to grow. Initially designed as simple token carriers, such as telephone cards, they first became secure document carriers (health card) before becoming mobile code carriers (Java Card). The major application areas of microprocessor cards are [5]:

- Telecommunications: with SIM cards for GSM mobile networks, and telephone cards
- Banking: with EMV payment cards and electronic wallets such as Moneo
- Security and access control: with electronic identity cards, biometric passports, pay-per-view television, and contactless payment with the RATP's Passe Navigo card
- Health: with the French Vitale card and the future NetCards card in Europe.

4. Standards and characteristics

4.1 Standardization level of smart cards

The level of standardization of the smart card is remarkable. Whether it is a bank card or a SIM card, it will be recognized by the reader device (mobile phone or bank ATM). Three types of parameters are standardized: physical parameters that set the size and positions of the chip and its contacts, electrical parameters specifying the supply voltages, various pins, and software parameters defining the communication mode with the card.

The standardization of smart cards has resulted in the publication of international standards. The classic or contact smart card is standardized by the ISO7816 standards (-1,-2,-3, and -4). These standards define respectively, the physical characteristics of the card, the position and pinouts, the electrical levels, and the various basic commands [6]. On the other hand, contactless cards are governed by the ISO14443 standards (-1,-2, and -3). They contain specifications for the part, the electrical interface, as well as the communication and collision management protocol [7].

4.2 Different types of smart cards

The cards are divided into two families: contact cards (memory or microprocessor) and contactless cards [8].

4.3 Contact cards

A contact smart card has eight visible connectors. Three connectors are reserved for future use (RFU: Reserved for Future Use). The electrical power supply (pins VCC and VSS, usually 5 V) and the clock at around 3.5 MHz (Clock) are included. The smart card can be rebooted by the reader by briefly setting the RESET pin to 0 (hot reset). Communication between the chip and the reader can be in serial mode, bit by bit, on the input/output pin. The Vpp pin was previously used for programming the chip (**Figure 1**). There are two types of contact smart cards: memory cards and microprocessor cards [9].

4.3.1 Memory cards

This type of card consists essentially of an EEPROM memory that does not require high programming voltage. This memory is generally programmable only once (OTPROM: One Time Programmable). Memory cards are used in the field of telephony where the programming principle is irreversible.

4.3.2 Microprocessor cards

This type of smart card is composed of a chip used to perform complex functions. It can be considered as a mini-computer that includes all the components that are

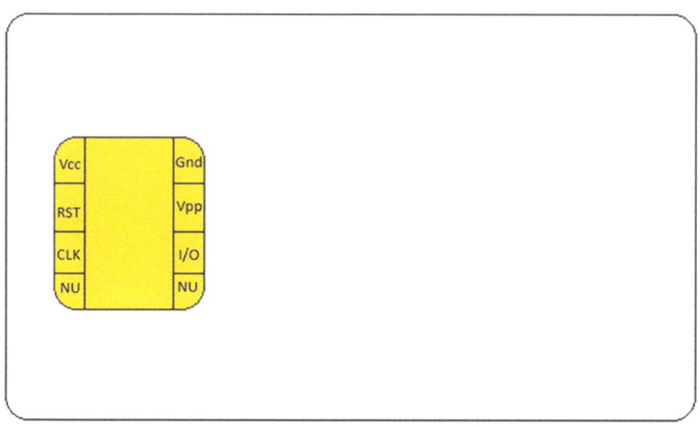

Figure 1.
Pinout of a contact smart card.

usually found on a PC motherboard on a system-on-chip (SoC). This chip includes a microprocessor (8-bit, 16-bit or 32-bit), a memory area (ROM, EEPROM, RAM), as well as several calculation devices used, among others, for cryptography (such as RSA, DES, Random, etc.) and a data transmission interface (UART). This chip has a surface area of less than 25 mm^2. They are particularly used in bank cards, health insurance cards, but also SIM cards (Subscriber Identity Module) used in mobile phones.

4.4 Contactless cards

Contactless smart cards or RFID (Radio Frequency Identification Device) cards are not directly connected to the reader by a physical contact; the connection is made through an electromagnetic field. To function, the contactless card must be placed at a distance of less than 3 cm from the reader. To be powered, the card uses inductive or capacitive coupling. The clock used for card synchronization can be internal, and the inputs/outputs are made by modulation of the power supply. This type of card is used for access control systems, animal identification, containers, consumer products, etc.

4.5 Combined cards (Combi)

Combi cards, also referred to as dual interface cards, are cards that integrate contact and contactless technologies onto the same chip. They possess two distinct interfaces that enable their use in both contact and contactless modes, making them versatile and ideal for various applications like access control, public transportation, and electronic payment systems. The contact interface provides high security, while the contactless interface offers convenience and faster usage. Due to their ability to provide a seamless transition between contact and contactless modes, the use of combined cards is increasingly prevalent.

5. Steps of manufacturing

The manufacturing process of the electronic chip follows the same process as an integrated circuit, starting from the design and development phase, to the extraction of the wafer using specific CAD tools. These tools are used to make etchings on the wafer to produce the intended functionality of the chip. After the testing phase, the sawing process and finally the extraction of the chips are initiated [10].

The assembly of the chip onto the metal part of the card is illustrated in **Figure 2**. The legs of the chip are bonded to those of the protective module (side A) using a low-resistance wire. Side B represents the external contact support of the chip ensuring the connection with the reader.

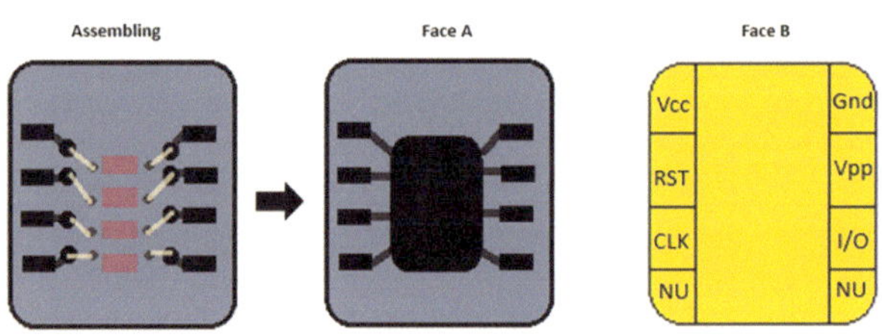

Figure 2.
Chip module realization.

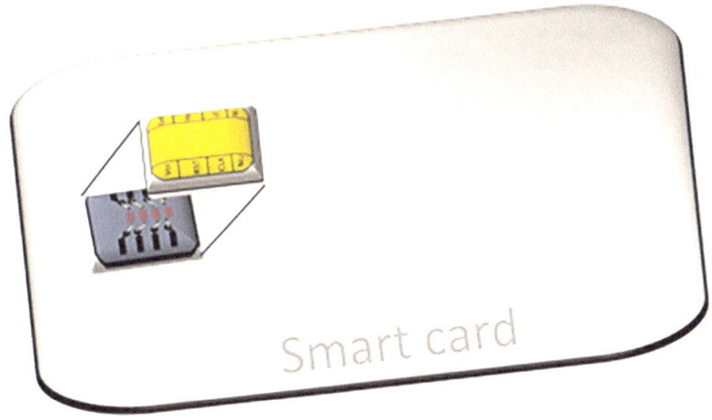

Figure 3.
Assembly processes for smart cards.

The chip module is then inserted into the PVC card. A cavity is excavated to fix the chip module (**Figure 3**).

Pour the manufacturing of the contactless smart card, the steps are similar to the contact smart card. However, an antenna is needed to facilitate data exchange by chemical etching of copper or aluminum on the PVC card.

6. Industrialized smart cards

The smart card industry has been constantly evolving in recent years thanks to the advancement of semiconductor technology as well as the widespread use of smart cards in modern society, such as in e-commerce, telecommunications, identification, access control, health, banking, entertainment, and transportation, etc.

The architectures proposed by different manufacturers depend on the application domains to which they are dedicated. The processor speed and memory block sizes (ROM, RAM, and EEPROM) integrated into the card's chip vary from one manufacturer to another. Industrialized cards may include 8/16/32-bit RISC architecture processors, clocked at frequencies ranging from 4 MHz for older versions up to 60 MHz for recent versions (**Table 1**) [11, 12].

Secure smart cards also exist in the global market with the integration of cryptographic modules ensuring the security of information stored in the chip as well as securing transmitted data.

7. Proposed smart card IP

The architecture consists of a 32-bit microprocessor with a 5-stage pipeline, which forms the core of our IP, memory blocks (ROM, RAM, EEPROM), a 32-bit crypto-processor (ECDSA, AES, RNG, SHA), and a communication interface (UART) connected to the card contact, ensuring communication with the reader (**Figure 4**).

During the design of the proposed smart card, standard mechanical and electrical constraints are taken into consideration [1]:

- Hardware Configuration: 32 KB ROM (for OS), 64 KB EEPROM (machine codes, information), and 8 KB RAM (data).

Manufacturer	Year	ROM (ko)	E^2PROM (ko)	Flash (ko)	RAM (ko)	Processor
Motorola SC01/ MAM01	1983	1,6	1 (EPROM)	NA	36o	6805, 8bits, 4 MHz
TRT-Philips P83C852	1995	16	4	NA	256o	80C51, 8 bits, 10 MHz RSA/DSA/ DES
Atmel AT90SDC100	2010	128	36	NA	6	AVR, 8/16 bits RISC, biprocessor, 30Mhz 3DES/AES/TRNG
Infineon SLE78CX1440P	2010	288	144	NA	8	Dual 16 bits, 33 MHz RSA/EC/AES/ 3DES/SHA2/TRNG
ST Microelectronics ST33J2M0	2016	NA	NA	2048	50	ARM, 32-bit RISC, 60 MHz AES/ DES,/Nescrypt co-processors
Samsung S3FT9MH_ID	2019	40	NA	500	14	SecuCalm16-bit 3DES/AES/RSA/ ECC

Table 1.
Characteristics of some industrialized smart cards.

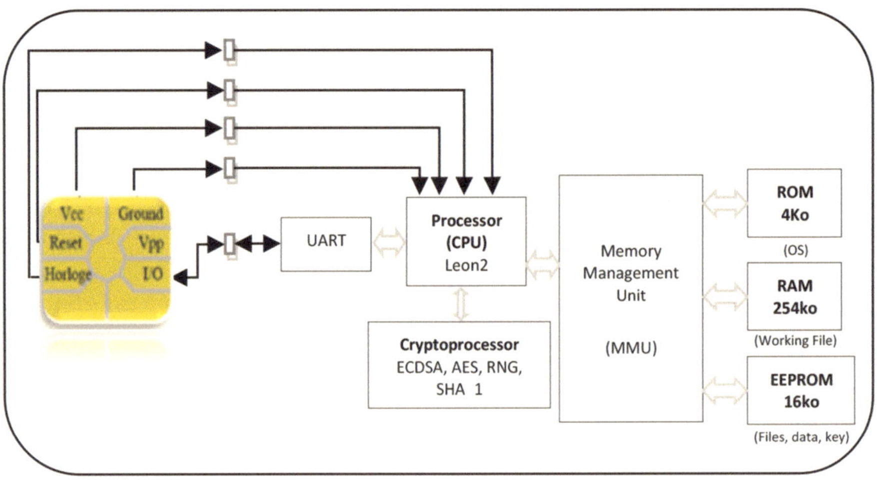

Figure 4.
Proposed IP smart card architecture.

- ISO7816-1/2 Standards: surface area of 25 mm^2, thickness of 200 µm, and dimensions of 85x54x0.76 mm.
- ISO7816-3 Standard: frequency range between 1 and 5 MHz for contact cards and 13.56 MHz for contactless cards according to the ISO14443 standard.

7.1 Choice of the processor

Currently, there is a wide range of dedicated processors for smart cards in the semiconductor market. Among this wide range, we can mention the AVR processor from Atmel, SecuCalm16 from Samsung, sc300, Cortex M0/3, and ARM7TDMI from ARM [13], the ST22XJ64 from STMicroelectronics, and SLE 88CX720P from Infineon Technologies [11–19].

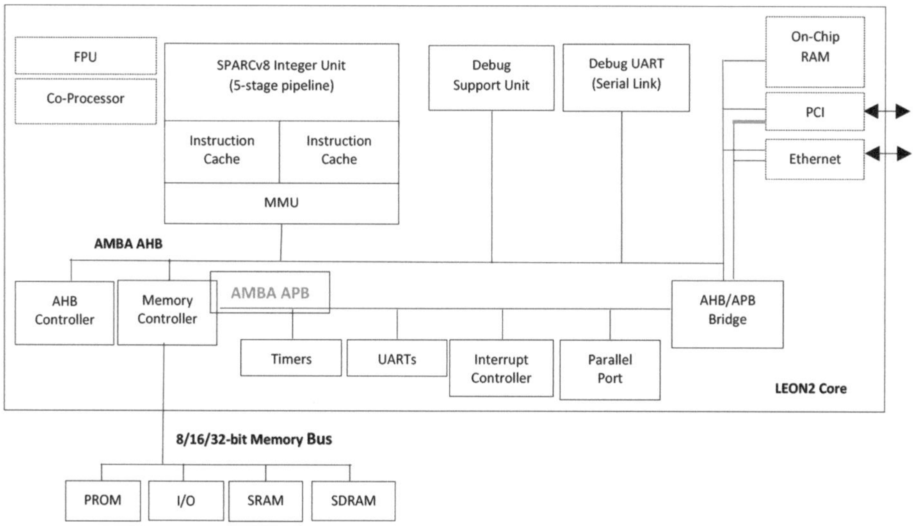

Figure 5.
LEON2 processor architecture.

In recent years, Gaisler Research, under contract with the European Space Agency (ESA), developed a processor called LEON2 [20]. It is used in embedded systems on board satellites. The LEON2 processor is available in two versions: Standard and Fault-tolerant. This processor is defined by a freely usable IP described in VHDL RTL (**Figure 5**).

The interest in the Leon2 processor is demonstrated by the recent production by ATMEL of a component for space applications based on this processor [19]. It is also used in an increasing number of applications thanks to its characteristic of being a freely usable IP. LEON2 was the core of an identification portable system described in [21, 22], and it has also been used to manage a wireless communication application in [23].

Apart from the license, several characteristics led to the choice of LEON2 as the core of our application:

- 32-bit RISC architecture with a 5-stage pipeline;
- Configurable instruction/data cache memories;
- A wide variety of possible configurations;
- Easy porting to different hardware targets;
- A UART interface for data transmission and reception;
- A memory controller that can reach 32-bit addresses and data;
- Possibility of adding new applications through its coprocessor.

7.2 Cryptographic mechanisms

An electronic smart card, whether it is a SIM card, credit card, access card, transportation card, or health card, includes a microprocessor that can store sensitive information. Hence the need to integrate security mechanisms and cryptographic means to ensure that information has not been altered during communication (integrity), to avoid disclosure of their content to third parties (confidentiality), and to identify the author of a document or transaction (authentication).

Modern electronic systems such as smart cards increasingly incorporate components called IPs that can be inserted into any type of design and which provide certain functionalities whose complexity can reach the heart of the processor. The data path of the chosen LEON2 processor is 32-bit, hence the need to adapt the different cryptographic modules. In this work, four IPs providing cryptographic mechanisms are proposed: SHA_1, RNG, ECDSA, and AES. The architectures of the developed IPs are 32 bits. Constraints related to the smart card are taken into consideration during the design of these IPs, which are speed, surface area, and power consumption.

7.2.1 Integrity mechanism

Integrity is a technique used to preserve the integrity of information. It is ensured by a function called hashing, which generates a fingerprint (or hash) of a message. The main functionality of hashing is to verify that the message received by the recipient has not been altered during transmission. This mechanism is also used for digital signature of the message. In this work, the SHA (Secure Hash Algorithm) hashing standard was used. It was designed by the United States National Security Agency (NSA) and published by the National Institute of Standards and Technology (NIST) in 1993. The SHA_1 hashing algorithm generates a 160-bit compressed output from a message of length less than 264-bits [24] with a block size of 512 bits. The first step of this algorithm is to fill or add (Padding) bits to the message M in such a way that the length of the resulting message is a multiple of 512 bits. Then, 80 logical functions defined on words and 80 32-bit constants are performed. These functions produce 32-bit words as output and take three 32-bit words as input [24]. The SHA_1 hashing function is described by a 32-bit architecture as follows (**Figure 6**).

The message to be hashed is loaded in blocks of 32-bits through the input interface. Then, the words are processed on 32-bits. Finally, an output interface generates 5 blocks of 32-bits constituting the hash (160-bits).

7.2.2 Random number generator mechanism

Secure random number generation (RNG) is an essential function in cryptography and for computer security in general. Cryptographic mechanisms are public, and their security is based on the secrecy of the encryption key (Kerckhoffs' principle). This key must be unpredictable and generated automatically to prevent the possibility of disclosure by an unauthorized third party. Modern cryptographic systems, such as digital signatures, rely heavily on random number generators for producing encryption keys [25].

A random number generator takes an input value, called a seed, and produces an output number that is the result of a computational algorithm. These functions are generally resource-intensive and time-consuming. Pseudo-random generators involve

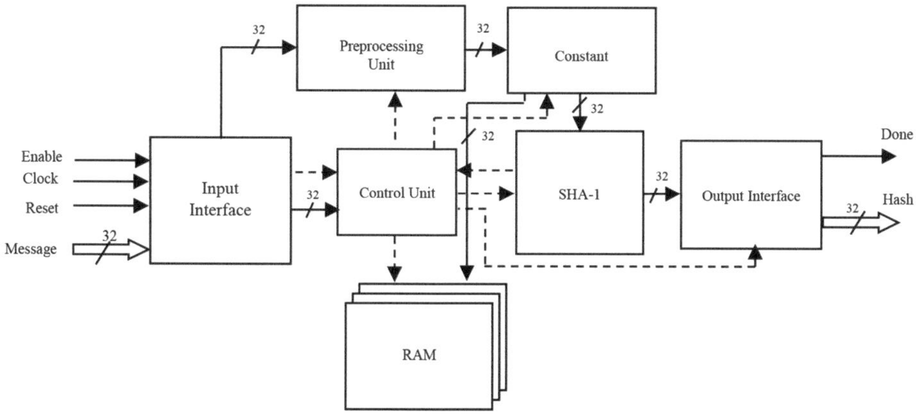

Figure 6.
32-bit SHA-1 function architecture.

applying a non-linear function by combining several linear feedback shift registers (LFSR) of different sizes.

In this work, the pseudo-random generator W7, which is a standard for GSM communication, was used for key generation due to its performance (speed, complexity, and low power consumption). It is a stream cipher algorithm published in April 2002 by Thomas, D. Anthony, T. Berson, and G. Gond. The internal architecture of W7 consists of three Linear Feedback Shift Registers (LFSR) of respective lengths 38, 43, and 47 bits with periods of 2^38-1, 2^43-1, and 2^47-1. Modifications were made in this manuscript to generate keys of size 163 bits to adapt to the datapath of LEON2. Specifically, each register was subdivided into two 32-bit registers. An output interface is used to group the randomly generated bits into 6 blocks of 32 bits to generate a random key of 164 bits (**Figure** 7).

7.2.3 Authentication mechanism

Operations such as bank transactions, personal authentication, and access to workplaces require the signature of the concerned person. Especially, when they are conducted via an open system like the internet. Hence arises the need to design digital signature mechanisms for the authentication of individuals and companies making purchases or sales over the internet.

Digital signature seeks to digitally mimic a handwritten signature. It consists of a string of bits that depends on the message and a secret key known only to the signer. In practice, digital signature schemes use a hash function that generates a digital fingerprint of a message m to be signed.

There are several digital signature standards that have been developed, such as DSA (Digital Signature Algorithm), digital signature based on the RSA algorithm (Rivest, Shamir, and Adleman), and digital signature based on elliptic curves (ECDSA) which appear in standards ANSI X9.62, FIPS 186-2, IEEE 1363-2000, and ISO/IEC 15946-2 [26]. This scheme, known to be safe and efficient for data authentication, has been used since 2000 by many banks for customer authentication, having key sizes of the order of 163, 271, and 571 bits. It is dedicated to support with specific constraints for smart cards. The 32-bit architecture of the ECDSA digital signature scheme is illustrated by **Figure 8**.

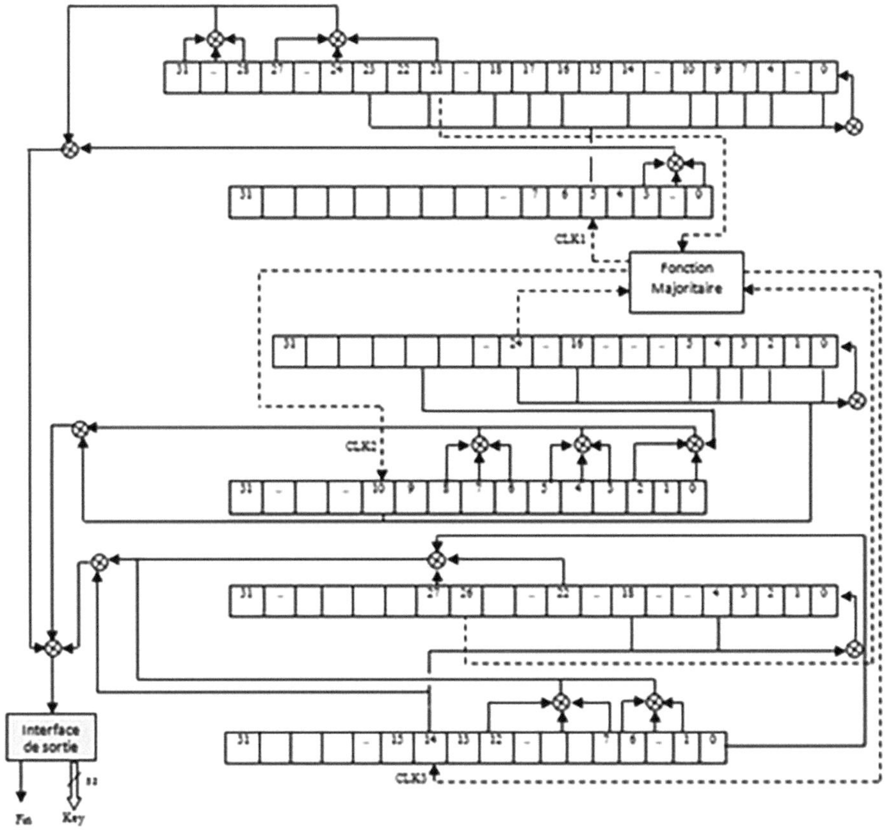

Figure 7.
32-bit W7 random number generator architecture.

The proposed architecture consists of a SHA_1 hash block, a random number generator (RNG), a library of arithmetic operators over the Galois field GF(2n), and modular operations (inversion, multiplication, and addition) based on 32-bit architectures. This library is necessary to perform operations on elliptic curves as well as scalar multiplication KP, which represents the basic operation for the elliptic curve digital signature algorithm (ECDSA). The ENABLE signal initiates data input. The clock signals CLK and RESET enable the block to be synchronized, and the DONE signal indicates the end of the operation. A control unit is responsible for activating/deactivating the key pair generation process, as well as signature generation/verification.

7.2.4 Confidentiality mechanism

The confidentiality mechanism ensures that information is made unintelligible to unauthorized individuals, entities, and processes.

In this work, the AES (Advanced Encryption Standard) algorithm is chosen to secure data stored in the smart card. It is a block cipher encryption/decryption algorithm, where messages are encrypted in blocks of 128 bits (16 bytes) with key sizes of 128, 192, or 256 bits. The key size defines its level of security, with larger key sizes

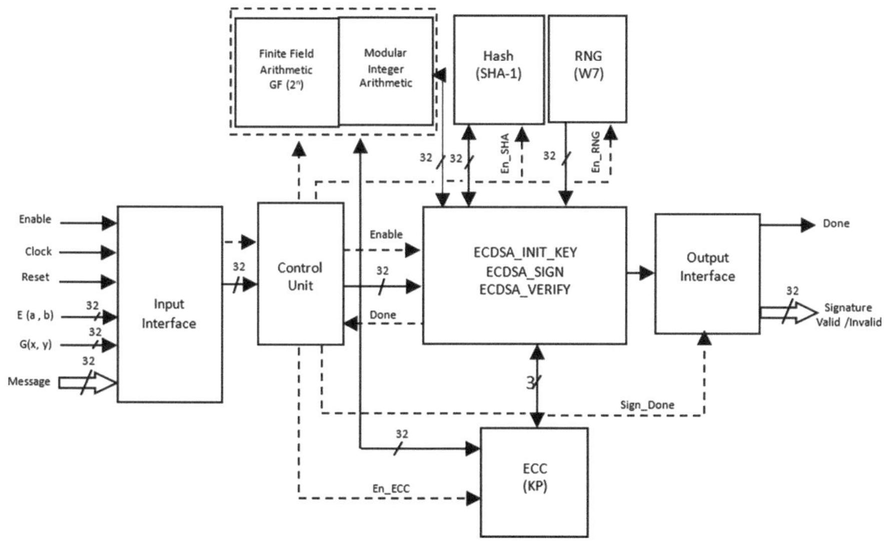

Figure 8.
32-bit ECDSA based digital signature architecture.

providing higher security levels [27]. This algorithm has been chosen to be fully operational and secure in any type of environment, which encouraged us to opt for AES with a 128-bit key to ensure the confidentiality service of the smart card. The choice of this algorithm meets many criteria such as its robustness against potential attacks, high processing speed, low resource and memory requirements, and ease of implementation (SP Network) with great flexibility.

In the remainder of this chapter, we have chosen AES with a 128-bit key in its version with 10 rounds. Initially, the plaintext is combined with the first round key K0, equal to the key, through the ADDRoundKey function. Each of the first nine rounds consists of four transformations: SubBytes (4 32-bit SB (i) blocks), ShiftRows, MixColumns, and ADDRoundKey. The last round consists of the same functions as a regular round, except for the MixColumns transformation.

The 32-bit architecture of the AES algorithm is described by **Figure 9**. An Input_Buffer input buffer allows loading the message to be encrypted (or decrypted) in 32-bit blocks. The 128-bit key is loaded onto 4 32-bit blocks. For decryption, the InvSubBytes (4 32-bit InvSB (i) blocks), InvShiftRows, and InvMixColumns functions are used. A control unit manages the activation and deactivation of the different blocks. An Output_Buffer output buffer allows the encrypted message to be returned in 4 32-bit blocks.

8. Modified architecture of Leon2 processor

Modifications were made to the block diagram of the Leon2 processor by adding a crypto-processor and external memories (PROM, SRAM, and EEPROM). These different blocks were developed in the previous sections. Unlike software implementation, which suffers from poor performance, the hardware implementation of cryptographic primitives is recognized to be more efficient in terms of speed, memory usage, and power consumption for embedded systems.

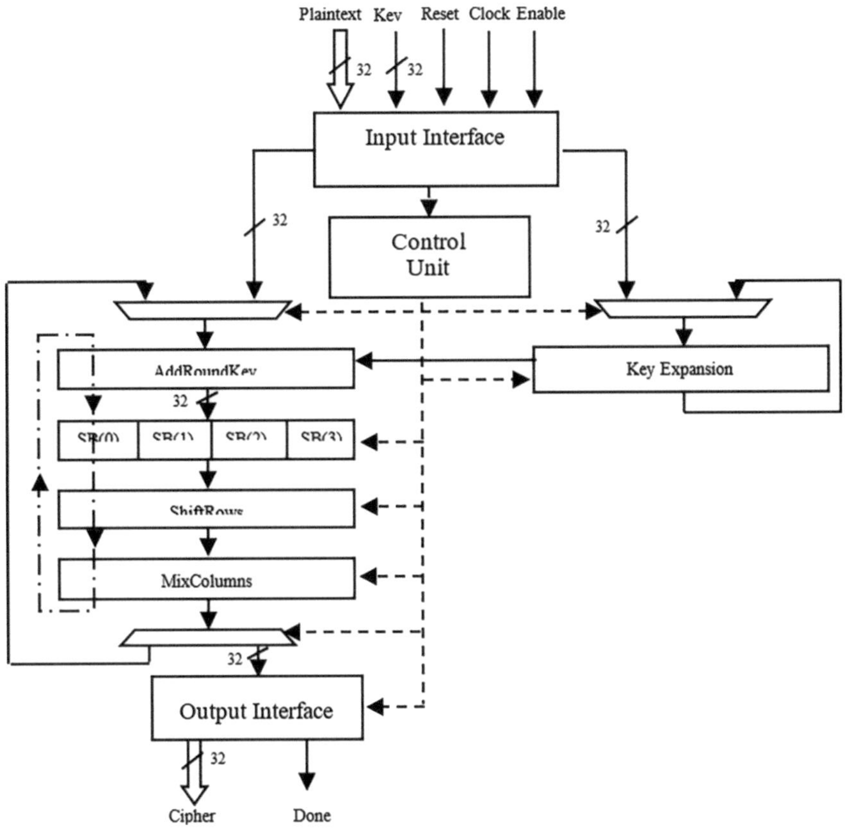

Figure 9.
32-bit architecture of encrypt/decrypt AES algorithm.

8.1 Cryptographic instructions set extension of the LEON2 processor

A variety of work is focused on improving the security of processors and aims to extend cryptographic instructions. The work described in [28, 29] presents sets of elliptic curve instructions that have been incorporated into a variety of processors. In publications [30–33], the authors focused on extending instructions for a random number generator, cryptographic modules, symmetric cryptosystems, and the AES algorithm. Hardware implementation of cryptographic primitives is recognized to be more efficient compared to software implementation, in terms of speed, occupied surface, and power dissipated for embedded applications. Hardware implementation also ensures a higher level of security, as a circuit cannot be easily attacked. Hence, there is an interest in hardware implementation in the continuation of our work.

In this section, we will present the principle of integrating new instructions into the core of the LEON2 processor to support the developed cryptographic mechanisms (RNG, SHA, ECDSA, and AES). To achieve this, the entire unit (IU) of the LEON2 is extended by integrating cryptographic instructions through coupling hardware IPs to the processor's data path to extend its instruction set (See **Figure 10**).

There are several return paths from the different stages to the decoding and execution stages. The memory stage is connected to the data cache. The cryptographic unit (CU) grouping cryptographic primitives extensions is described in previous

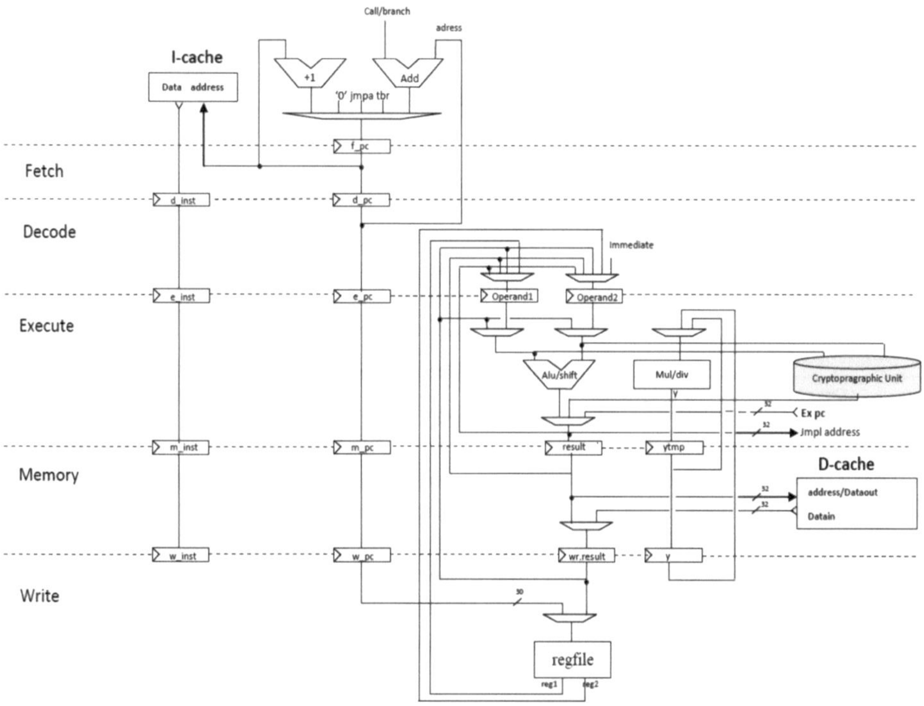

Figure 10.
Modified architecture of Leon2 integer unit.

sections. The cryptographic unit is implemented in parallel with the "ALU/Shifter" unit. The operands "op1" and "op2" of the cryptographic unit can be blocked at the input of the ALU/Shifter, as long as this unit is active. There are two input registers "rs1.data" and "rs2.data". Additional multiplexers on the return paths and output paths prevent the propagation of critical data.

8.2 Proposed instructions

The LEON2 processor is a 32-bit SPARC V8 RISC architecture that has different instruction formats with three and two inputs and one output operand.

8.2.1 SHA_1 instruction

The SHA_1 instruction consists of a 32-bit operand rs1 for inputting the message to be hashed in 32-bit blocks. The result is stored in the destination register rd on 32 bits. The principle of this instruction is shown in Figure II.22. The format of this instruction is as described by Eq. (1).

$$SHA - 1\, rs1\, rd \tag{1}$$

8.2.2 RNG instruction

The RNG instruction is a hardware instruction used to generate random numbers, and it does not take any operands. The result of this instruction, a 32-bit random

[20] National Aerospace Laboratory NLR. On-board Payload Data Processing for SAR and Multispectral data processing on-board satellites (LEON2/FFTC). In: The 2nd International Workshop on On-Board Payload Data Compression, 28–29 October 2010. Toulouse; 2010

[21] Hwang D, Schaumont P, Fan Y, Hodjat A, Lai B, Sakiyama K, et al. Design flow for HW/SW Accelleration transparency in the Thumbpod secure embedded system. In: 40th Design Automation Conference (DAC 2003), 2–6 June 2003. Anahiem, CA; 2003. pp. 60-65

[22] Stamenkovic Z, Wolf C, Schoof G, Gaisler J. LEON-2: General Purpose Processor for a Wireless Engine. IEEE Design and Diagnostics of Electronic Circuits and systems. 2006;**2006**: 48-51

[23] Charoenpanyasak S, Suntiamorntut W. The next generation of sensor node in wireless sensor networks. Journal of Telecommunications. 2011;**9**:48

[24] National Institute of Standards and Technology (NIST). Secure Hash Standard, FIPS PUB 180-1, 2002

[25] National Institute of Standards and Technology (NIST), Gaithersburg, Maryland, Special Publication 800-57: Recommendation for Key Management. Part1: General Guideline. 2003. Available from: http://csrc.nist.gov/CryptoToolkit/tkkeymgmt.html

[26] ANSI. Public Key Cryptography For The Financial Services Industry: The Elliptic Curve Digital Signature Algorithm (ECDSA). ANSI X9.62. 2005

[27] Bur WE. Selecting the advanced encryption standard. IEEE Security & Privacy Magazine. 2003;**1**(2):43-52

[28] Cohen AE, Parhi KK. Fast reconfigurable elliptic curve cryptography acceleration for GF (2m) on 32 bit processors. Journal of Signal Processing Systems. 2010;**60**(1):31-45

[29] Bartolini S, Branovic I, Giorgi R, Martinelli E. Effects of instruction-set extensions on an embedded processor: A case study on elliptic curve cryptography over GF(2^m). IEEE Transactions on Computers. 2008;**57**(5):679-685

[30] Drutarovsky M, Varchola M. Cryptographic system on a chip based on soft-core with embedded true random number generator. In: Proceedings of the 11th IEEE Workshop on Design and Diagnostics of Electronic Circuits and Systems (DDECS '08). Washington, DC, USA: IEEE Computer Society; 2008. pp. 164-169

[31] O'Melia SR, Elbirt AJ. Instruction Set Extensions for Enhancing the Performance of Symmetric-Key, Cryptography, 2008 Annual Computer Security Applications Conference (ACSAC). 2008. pp. 465-474

[32] Glenn H, Parviz P, Michael T, Gardner JS, Bryce A, Jim D, et al. High-performance deep-learning coprocessor integrated into x86 SoC with server-class CPUs. In: Proceedings of the ACM/IEEE 47th Annual International Symposium on Computer Architecture (ISCA'20). IEEE Press; 2020. pp. 15–26

[33] O'Melia SR, Elbirt AJ, Enhancing the Performance of Symmetric-Key Cryptography via Instruction Set Extensions. IEEE Transactions on Very Large Scale Integration (VLSI) Systems. 2010;**18**:1505-1518